Holt McDougal Mathematics

Course 2
Chapter 1 Resource Book

ISBN 13: 978-0-55-400598-0
ISBN 10: 0-55-400598-0

3 4 5   1417   12 11 10

4500256712

Contents

Holt McDougal Mathematics

Description of Contents

Family Involvement Pages

The Chapter Resource Book includes a set of family involvement pages for each section. The family involvement pages consist of the following items:

- **Family Letter**, a two-page letter describing the math that the student will study in each section. A list of vocabulary words from the lessons is included, and some worked-out examples are provided.
- **At-Home Practice**, a one-page worksheet with problems drawn from the content described in the Family Letter. Answers are provided on the page so that the student and his or her family can check this work at home.
- **Family Fun**, a one-page activity sheet that the student and his or her family can work on together.

Practice A, B, and C

There are three practice worksheets for every lesson. All of these reinforce the content of the lesson. Practice B is shown in the Teacher's Edition and is appropriate for the on-level student. It is also available as a workbook (Homework & Practice Workbook)

Practice A is easier than Practice B but still practices the content of the lesson. Practice C is more challenging than Practice B.

Review for Mastery

The Review for Mastery worksheet (one per lesson) provides an alternate way to teach or review the main concepts of the lesson. This worksheet is one or two pages long and is shown in the Teacher's Edition.

Challenge

The Challenge worksheet (one per lesson) enhances critical thinking skills and extends the lesson. This worksheet is shown in the Teacher's Edition.

Problem Solving

The Problem Solving worksheet (one per lesson) provides practice in problem solving and opportunities for real-world applications and for interdisciplinary connections. There are both multiple choice and short response problems. This worksheet is shown in the Teacher's Edition.

Reading Strategies

The Reading Strategies worksheet (one per lesson) provides tools to help the student master math vocabulary or symbols.

Puzzles, Twisters & Teasers

The Puzzles, Twisters & Teasers worksheet (one per lesson) provides fun practice while reinforcing the content of the lesson.

Family Letter

1A Patterns and Relationships

Dear Family,

The student is learning to identify and extend number patterns. He or she is also learning to represent numbers by using exponents. Just as multiplication is valuable when you are representing repeated addition, exponents are valuable when you are representing repeated multiplication. An **exponent** is a part of a **power**. It is a number that tells how many times the **base** is to be multiplied by itself. A power with base 3 and exponent 4 is written 3^4. The example below shows how the student will find the value of a number with an exponent.

Find the value of 3^4.

$3^4 = 3 \cdot 3 \cdot 3 \cdot 3 = 81$ Multiply the base 3 by itself 4 times.

The student will use powers of ten when expressing numbers with **scientific notation**. To multiply by a power of ten with positive exponents, move the decimal point to the right the same number of places as the value of the exponent.

Multiply $258 \cdot 10^4$.

Remember: $258 = 258.0 = 258.0000$

$258 \cdot 10^4 = 258.0000$ Move the decimal point 4 places to the right. Add

$= 2{,}580{,}000$ 4 zeros as placeholders.

Write 45,250,000 in scientific notation.

4.5250000   Move the decimal point left to form a number that is greater than or equal to one but less than 10.

10^7 Since the decimal point was moved 7 places to the left, the power of ten is 7.

4.525×10^7 Write the number in scientific notation.

Note that $4.525 > 1$ but < 10.

As the student works with more complex numbers, he or she will continue to explore the use of scientific notation.

Vocabulary

These are the math words we are learning:

Associative Property property that states that when you add or multiply, you can group numbers in any combination

base when a number is written in exponential form, the number that is used as a factor is the base

Commutative Property property that states that when you add or multiply, you can do so in any order

conjecture a statement believed to be true

Distributive Property property that states that a number times a sum equals the sum of the products of that number and each addend

exponent a number that tells how many times to use the base as a factor

Identity Property property that states that the product of 1 and a number and the sum of 0 and a number is that number

numerical expression an expression made up of numbers and operations

Holt McDougal Mathematics

Family Letter

1A Patterns and Relationships continued

The student will follow the order of operations, a set of rules that standardize how to simplify expressions.

Order of Operations
1. Perform operations within grouping symbols.
2. Evaluate powers.
3. Multiply and divide in order from left to right.
4. Add and subtract in order from left to right.

Here is how the student will simplify an expression using the order of operations.

Simplify $3^2 \cdot 6 + 4$.

$3^2 \cdot 6 + 4$	There are no parentheses so evaluate the power.
$9 \cdot 6 + 4$	Multiply.
$54 + 4$	Add.
58	

The following properties will also help the student simplify expressions. These properties apply to all numbers.

Property	Words	Algebra
Commutative Property	You can add or multiply numbers in any order.	$a + b = b + a$ $k \cdot m = m \cdot k$
Associative Property	When you add or multiply, you can group the numbers together in any combination.	$(p + s) + t$ $= p + (s + t)$ $(x \cdot y) \cdot z$ $= x \cdot (y \cdot z)$
Identity Property	The sum of 0 and any number is the number. The product of 1 and any number is the number.	$x + 0 = x$ $x \cdot 1 = x$
Distributive Property	A multiplier outside the parentheses of another operation can be applied to each number inside the parentheses before the operation is performed.	$d(e + f)$ $= d(e) + d(f)$

Sincerely,

order of operations the rules that must be followed when simplifying expressions

power a number written as a base and an exponent

scientific notation a type of shorthand for writing numbers

Holt McDougal Mathematics

<table>
<tr><td>CHAPTER
1</td><td></td></tr>
</table>

At-Home Practice

1A Patterns and Relationships

Find each value.

1. 3^5

2. 9^4

3. 2^6

4. 11^3

5. 8^3

6. 5^4

7. 10^2

8. 1^9

Write each number using an exponent and the given base.

9. 16, base 2

10. 81, base 9

11. 27, base 3

12. 25, base 5

Multiply.

13. $12 \cdot 10^5$

14. $4 \cdot 10^3$

15. $27 \cdot 10^0$

16. $35 \cdot 10^4$

17. $428 \cdot 10^2$

18. $576 \cdot 10^5$

19. $8{,}791 \cdot 10^4$

20. $68{,}874 \cdot 10^1$

21. $1{,}783 \cdot 10^6$

22. $27{,}845 \cdot 10^3$

Write each number in scientific notation.

23. 728,000

24. 245,000

25. 8,000,000

26. 981.2

Simplify each expression using the order of operations.

27. $3 \cdot 7 + 7 + (6 - 4)^2$

28. $7{,}711 \div 11 - 25 \cdot 4$

29. $45 \div (3)^2$

30. $(84 \div 4) \cdot 3$

31. $(20 \cdot 2) - (12 \div 3)$

32. $6^3 \div 3 + 15$

Answers: 1. 243 **2.** 6,561 **3.** 64 **4.** 1,331 **5.** 512 **6.** 625 **7.** 100 **8.** 1 **9.** 2^4 **10.** 9^2 **11.** 3^3 **12.** 5^2 **13.** 1,200,000 **14.** 4,000 **15.** 27 **16.** 350,000 **17.** 42,800 **18.** 57,600,000 **19.** 87,910,000 **20.** 688,740 **21.** 1,783,000,000 **22.** 27,845,000 **23.** 7.28×10^5 **24.** 2.45×10^5 **25.** 8×10^6 **26.** 9.812×10^2 **27.** 32 **28.** 601 **29.** 5 **30.** 63 **31.** 36 **32.** 87

  Holt McDougal Mathematics

Family Fun
Shining Stars

Directions

Use any whole number, each of the four basic operations, and the order of operations to arrive at the total in the middle of the stars. Plot your answers on the points of the stars and move clockwise around the star, starting and ending at the apex of the star. Write the operation symbol being used in between each point.
The 4-star is done for you.

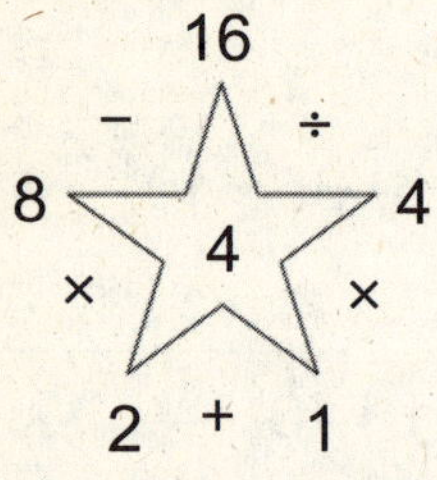

Use the numbers in the stars to come up with your own game that uses the order of operations. Give your creation to a friend and see if he or she can figure it out.

Possible answers: 10: 100 ÷ 10 × 1 + 5 × 20 − 100; 9: 3 × 3 + 100 ÷ 20 − 15 ÷ 3; 16: 2 × 8 + 8 + 8 − 32 ÷ 2; 2: 2 × 1 + 6 ÷ 3 − 4 ÷ 2; 0: 100 − 100 ÷ 1 + 4 × 0 × 100

Holt McDougal Mathematics

Name ____ Date ________________ Class ________________

Practice A
Numbers and Patterns

Write a rule that describes the pattern. Then write the next three numbers in the pattern.

1. 7, 10, 13, 16, _____, _____, _____, ...

   The pattern is

   ____

   ____

2. 81, 70, 59, 48, _____, _____, _____, ...

   The pattern is

   ____

   ____

3. 2, 4, 8, 16, _____, _____, _____, ...

   The pattern is

   ____

   ____

4. 1, 3, 6, 10, 15, _____, _____, _____, ...

   The pattern is

   ____

   ____

Write a rule that describes the pattern. Then draw the next three figures in the pattern.

5. 

   The pattern is ____

6. 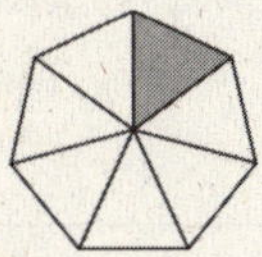 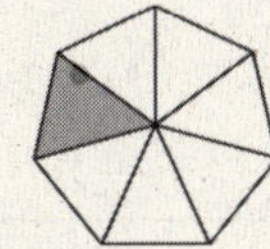 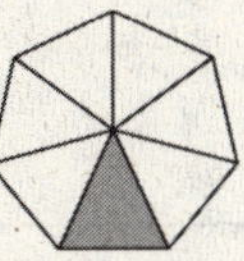 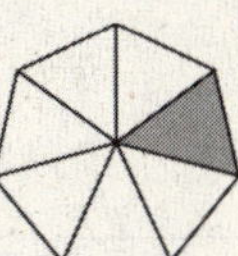

   The pattern is ____

7. Complete the table so that it shows the number of dots in each figure.

   Figure 1   Figure 2   Figure 3   Figure 4   Figure 5

Figure	1	2	3
Number of Dots			

   How many dots are in the fifth figure? _____

   Use drawings to justify your answer.

Holt McDougal Mathematics

LESSON 1-1

Practice B

Numbers and Patterns

Identify a possible pattern. Use the pattern to write the next three numbers.

1. 41, 37, 33, 29, _____, _____, _____, ...

   ____

   ____

2. 50, 52, 56, 62, _____, _____, _____, ...

   ____

   ____

3. 320, 160, 80, 40, _____, _____, _____, ...

   ____

   ____

4. 24, 40, 56, 72, _____, _____, _____, ...

   ____

   ____

Identify a possible pattern. Use the pattern to draw the next three figures.

5.

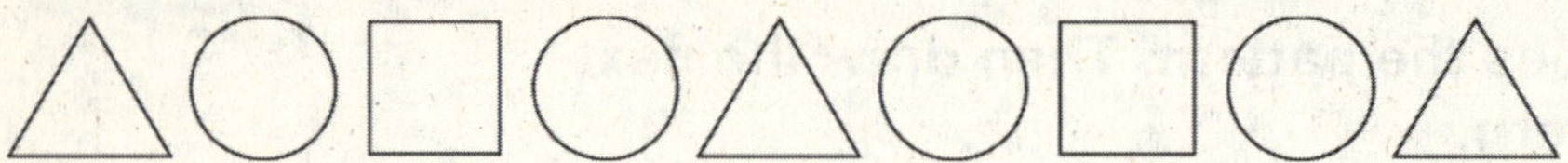

   ____

6.

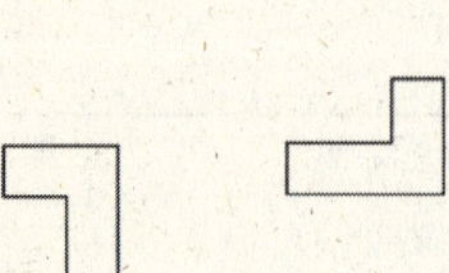  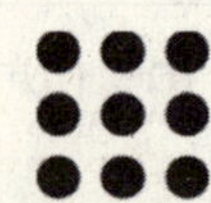

   ____

7. Complete the table so that it shows the number of dots in each figure.

Figure 1   **Figure 2**   **Figure 3**   **Figure 4**   **Figure 5**

Figure	1	2	3
Number of Dots			

How many dots are in the fifth figure of the pattern? _____

Use drawings to justify your answer.

     Holt McDougal Mathematics

LESSON 1-1	**Practice C**
	Numbers and Patterns

Identify a possible pattern. Use the pattern to write the missing numbers.

1. 72, 95, 118, 141, _____, _____, _____, ...

____

____

2. 65, _____, 51, 44, _____, _____, 23, ...

____

____

3. 4, 12, _____, 108, 324, 972, _____, ...

____

____

4. _____, _____, 47, 44, 40, _____, 29, ...

____

____

Look for a possible pattern. Use the pattern to draw the next three figures.

5. 

6. 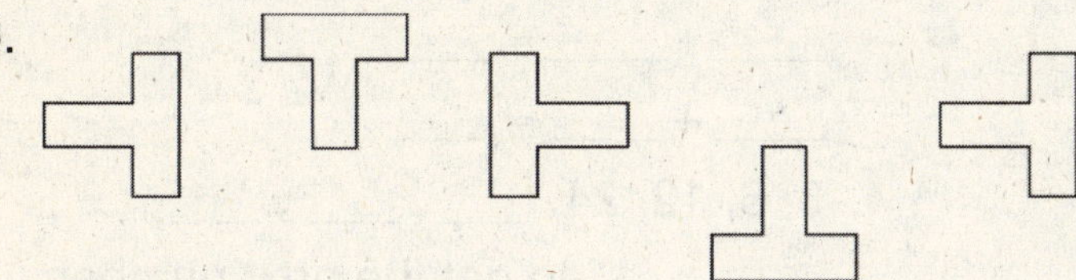

7. 

8. Complete the table so that it shows the number of squares in each figure.

 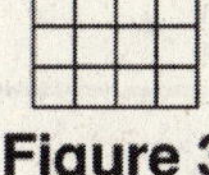

Figure 1   Figure 2   Figure 3   Figure 4   Figure 5

Figure	1	2	3
Number of Dots			

How many squares are in the fifth figure of the pattern? _____

Use drawings to justify your answer.

Holt McDougal Mathematics

| LESSON 1-1 | **Review for Mastery**
Numbers and Patterns |

To identify a number pattern, ask: What can I do to each number to get the number that comes next?

The pattern is to add 4 to get the next number. Use the pattern to get the next three numbers.

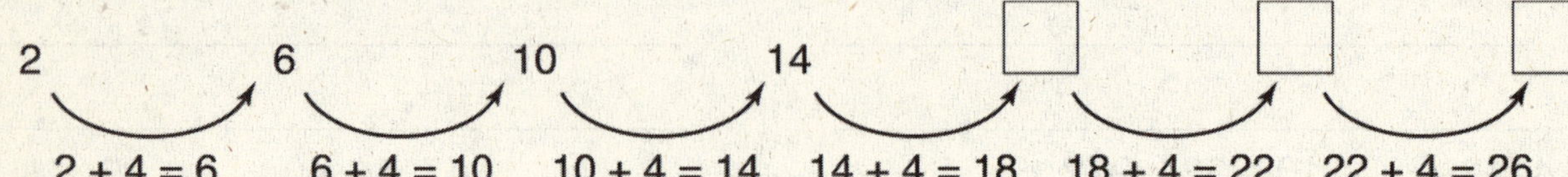

2 → 6 → 10 → 14 → ☐ → ☐ → ☐

$2 + 4 = 6$ $6 + 4 = 10$ $10 + 4 = 14$ $14 + 4 = 18$ $18 + 4 = 22$ $22 + 4 = 26$

and
$2 \times 3 = 6$

but
$6 \times 3 \neq 10$

Identify the pattern. Use the pattern to find the next three numbers.

1. 28, 25, 22, ____, ____, ____, ...

   Subtract ___ to get the next number.

   $22 - \underline{} = \underline{}$

   _____ − ___ = _____

   _____ − ___ = _____

2. 6, 24, 42, 60, ____, ____, ____, ...

   Add ____ to get the next number.

   $60 + \underline{} = \underline{}$

   _____ + _____ = _____

   _____ + _____ = _____

3. 4, 12, 20, 28, ____, ____, ____, ...

   __________ to get the next number.

4. 3, 6, 12, 24, ____, ____, ____, ...

   __________ to get the next number.

To identify a geometric pattern, ask: How can I change each figure to get the next figure?

The pattern is to shade the next square in a clockwise direction. Use the pattern to get the next three figures.

   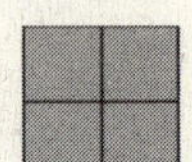 

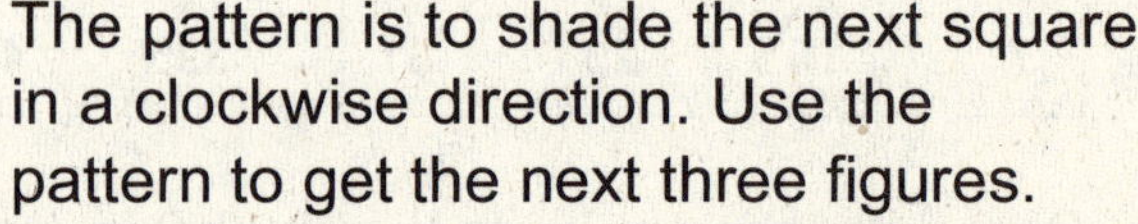

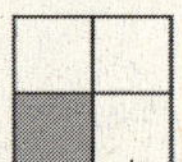 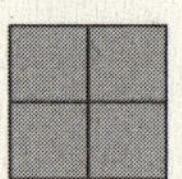

Draw the next three figures in each pattern.

5.
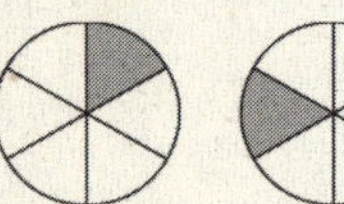   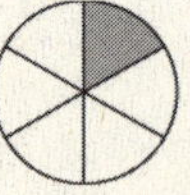 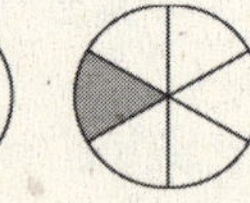

6.
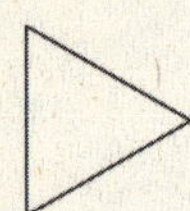 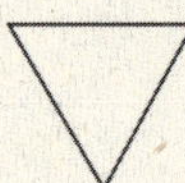 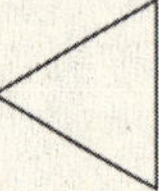 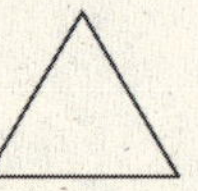 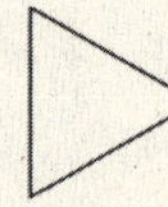

   Holt McDougal Mathematics

Challenge

LESSON 1-1

Pascal's Triangle

The pattern at the right is called Pascal's Triangle. This pattern is named after the French mathematician, Blaise Pascal, because he wrote about its properties.

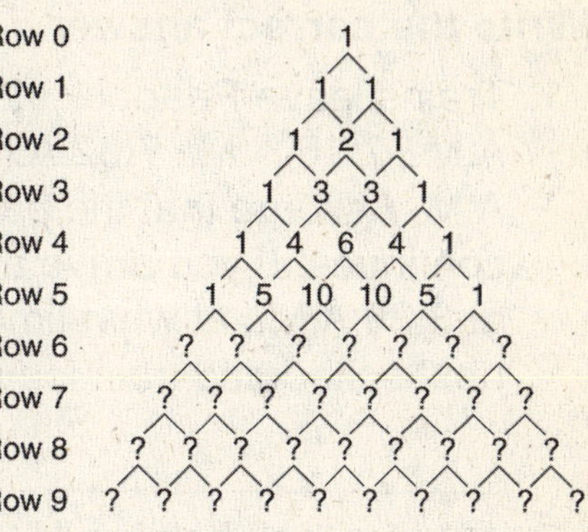

Use Pascal's Triangle to solve problems 1–5.

1. Write the sum of the numbers in each of rows 1–5. What pattern do you see?

   ____

   ____

2. Use the pattern that you found in problem 1 to predict the sum for row 9. _________

3. Look at the numbers in row 4 and the numbers they are connected to in row 3. Then look at the numbers in row 5 and the numbers they are connected to in row 4. What pattern do you see?

   ____

   ____

   ____

4. Use the pattern you found in problem 3 to find the numbers for rows 6–9.

   row 6: _________________________________

   row 7: _________________________________

   row 8: _________________________________

   row 9: _________________________________

5. Find the sum of the numbers in row 9. How does it compare to your prediction in problem 2? ________________________________

Problem Solving

LESSON 1-1

Numbers and Patterns

Write the correct answer.

1. Trains leave Peapack station at 7:00 A.M., 7:16 A.M., 7:32 A.M., and 7:48 A.M. Assume that the pattern continues. If you arrive at the station at 8:30 A.M., at what time will the next scheduled train leave?

   ____

2. Suppose the pattern 8, 16, 24, 32, ... is continued forever. Will the number 174 appear in the pattern? Why or why not?

   ____

3. A water tank holding 100 gallons begins to leak at 6:00 P.M. At 7:00 P.M. the tank has 94 gallons. At 8:00 P.M the tank has 88 gallons. At 9:00 P.M. the tank has 82 gallons. If the pattern continues, how many gallons will be left at 11:00 P.M.?

   ____

4. Awilda is making a necklace with sphere, pyramid, and cube shaped beads. If she continues the pattern below, what are the shapes of the next five beads?

Choose the letter for the best answer.

5. The drawing at the right shows the first three figures in a pattern. If the pattern continues, how many circles will be in the fifth figure of the pattern?

   A 20   C 27

   B 24   D 35

Figure 1   Figure 2   Figure 3

6. The table shows the number of handshakes if each person in a group shakes each other's hand once. How many handshakes will there be if there are 9 people?

   F 20   H 36

   G 30   J 45

People	2	3	4	5	6
Number of Handshakes	1	3	6	10	15

7. Melissa wrote the following number pattern: 71, 66, 59, 50, 39, ... What is the next number in the pattern?

   A 26   C 30

   B 27   D 35

8. By July 1, Bob saved $120. By July 8, he saved $185. By July 15, Bob saved $250. If the pattern continues, how much will he have saved by July 29?

   F $315   H $390

   G $380   J $415

          Holt McDougal Mathematics

Reading Strategies

LESSON 1-1

Identify Relationships

To identify and extend a **number pattern,** you must find the **relationship** between each number in the pattern and the number that it comes after.

4, 16, 28, 40, ☐, ☐, ☐, ...

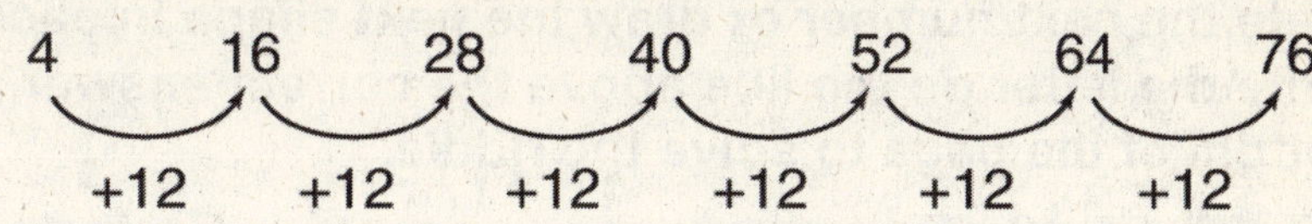

The relationship is that each number is 12 more than the number it comes after. So, the pattern is to add 12 to each number to get the next number.

To identify and extend a **geometric pattern**, you must find the **relationship** between each figure in the pattern and the figure that it comes after.

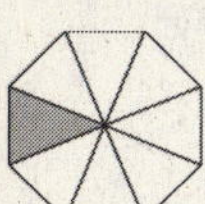 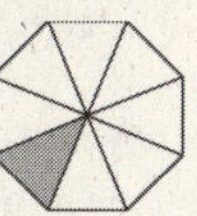 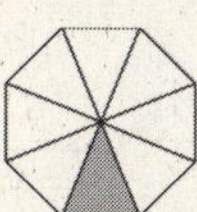 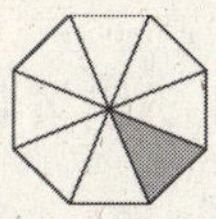   ?   ?   ?

The relationship is that the shaded triangle moves one triangle counterclockwise from the figure it comes after.

So, the next three figures in the pattern are:

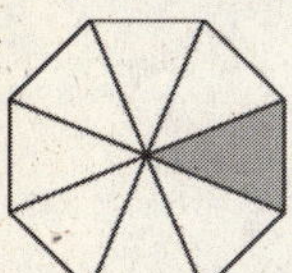 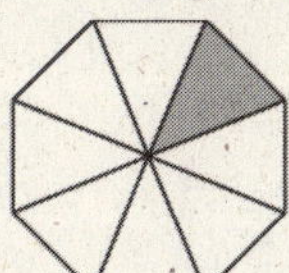 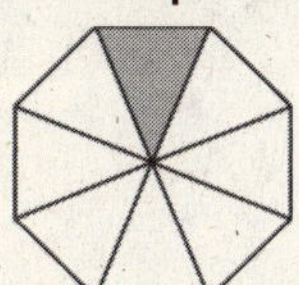

Use the pattern 64, 55, 46, 37, ☐, ☐, ☐, ... to answer questions 1 and 2.

1. What is the relationship between each number and the number it comes after?

____

2. What are the next three numbers in the pattern? ___________________

Use the pattern below to answer question 3.

  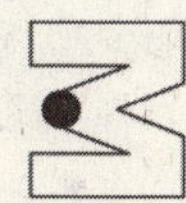  

3. What is the relationship between each figure and the figure it comes after?

____

     Holt McDougal Mathematics

LESSON 1-1 Puzzles, Twisters & Teasers

It's in the Cards!

When is it dangerous to play cards?

**Write the next number or draw the next shape in each pattern.
Write the letter on the line above the correct answer at the
bottom of the page to solve the riddle.**

T 75, 59, 43, 27, ____, ...

E 7, 14, 28, 56, ____, ...

H 12, 25, 39, 54, ____, ...

J 243, 81, 27, 9, ____, ...

W 28, 34, 42, 52, ____, ...

K 114, 91, 68, 45, ____, ...

R 2, 8, 32, 128, ____, ...

N 70, 56, 41, 25, ____, ...

S 5, 8, 14, 23, ____, ...

O 6, 30, 54, 78, ____, ...

L 6, 18, 54, 162, ____, ...

I 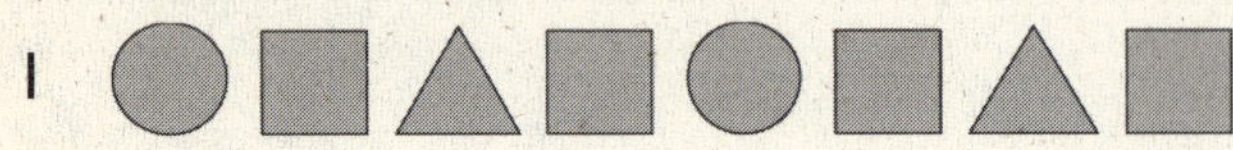

D 

<pre>
     ___ ___ ___ ___   ___ ___ ___
     64   70   112   8   11   70   112
</pre>

<pre>
___ ___ ___ ___ ___   ___   ___ ___ ___   ___
 3   102   22 112 512    35   64 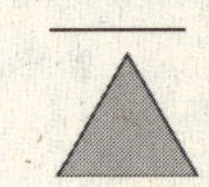 486   △
</pre>

  Holt McDougal Mathematics

<table><tr><td>**LESSON**
1-2</td><td></td></tr></table>

Practice A
Exponents

Multiply.

1. 4^2

    $4 \cdot 4 = $ ____

2. 2^3

    ____ $\cdot$ ____ $\cdot$ ____ = ____

3. 6^2

    ____ $\cdot$ ____ = ____

4. 9^2

5. 4^3

6. 3^5

7. 8^0

____

____

____

____

8. 10^2

9. 3^4

10. 9^1

11. 2^5

____

____

____

____

Use an exponent and the given base to write each number.

12. 25, base 5

13. 3, base 3

14. 8, base 2

15. 1, base 4

____

____

____

____

16. 81, base 9

17. 64, base 4

18. 64, base 8

19. 9, base 3

____

____

____

____

20. 36, base 6

21. 16, base 2

22. 27, base 3

23. 400, base 20

____

____

____

____

24. The first day, Jessie has $2. The second day, she has twice as much money as the first day. The third day, she has twice as much money as the second day. Write the amount of money she has on the third day in exponential form. Then write the amount in standard form.

____

25. Kevin runs 5 miles on Monday. The total number of miles he runs that week is 5 times the number of miles he runs on Monday. How many miles does Kevin run that week?

____

   Holt McDougal Mathematics

LESSON 1-2

Practice B
Exponents

Find each value.

1. 5^2

2. 2^4

3. 3^3

4. 7^2

5. 4^4

6. 12^2

7. 10^3

8. 11^1

9. 1^6

10. 20^2

11. 6^3

12. 7^3

Write each number using an exponent and the given base.

13. 16, base 4

14. 25, base 25

15. 100, base 10

16. 125, base 5

17. 32, base 2

18. 243, base 3

19. 900, base 30

20. 121, base 11

21. 3,600, base 60

22. 256, base 4

23. 512, base 8

24. 196, base 14

25. Damon has 4 times as many stamps as Julia. Julia has 4 times as many stamps as Claire. Claire has 4 stamps. Write the number of stamps Damon has in both exponential form and standard form.

26. Holly starts a jump rope exercise program. She jumps rope for 3 minutes the first week. In the second week, she triples the time she jumps. In the third week, she triples the time of the second week, and in the fourth week, she triples the time of the third week. How many minutes does she jump rope during the fourth week?

Holt McDougal Mathematics

LESSON 1-2

Practice C

Exponents

Find each value.

1. 3^8

2. 7^5

3. 50^2

4. 9^4

5. 30^3

6. 12^5

Compare. Write <, >, or =.

7. 7^2 ____ 48

8. 9^3 ____ 810

9. 6^4 ____ 10^3

10. 100,000 ____ 10^6

11. 2^7 ____ 6^3

12. 4^6 ____ 8^4

Write each number using an exponent and the given base.

13. 343, base 7

14. 625, base 5

15. 1,728, base 12

16. 225, base 15

17. 1,000,000, base 100

18. 1,225, base 35

19. 6,561, base 9

20. 2,187, base 3

21. 8,000, base 20

22. Jacob has 7 times as many postcards as Austin has. Austin has
7 times as many postcards as Angela has. Angela has 7 times
as many postcards as Samuel has. Samuel has 7 postcards.
Write the number of postcards Jacob has in both exponential
form and standard form.

                                                                 Holt McDougal Mathematics

LESSON 1-2
Review for Mastery
Exponents

The exponent tells you how many times to multiply the base by itself.

base ⟶ 3^4 ⟵ **exponent**

- To find 3^4, multiply the base (3) times itself 4 times

$$3^4 = 3 \cdot 3 \cdot 3 \cdot 3 = 81$$

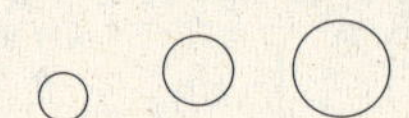

Find each value.

1. $4^3 =$ ____ • ____ • ____ $= 64$

2. $1^5 =$ • ____ • ____ • ____ • ____ • ____ $=$ ____

3. 5^2 4. 2^3 5. 3^3 6. 6^2

_______________   _______________   _______________   _______________

7. 8^2 8. 4^1 9. 5^3 10. 2^4

_______________   _______________   _______________   _______________

- You can write 64 using an exponent with the base 8.

$$8 \cdot 8 = 64$$
$$\text{So, } 64 = 8^2.$$

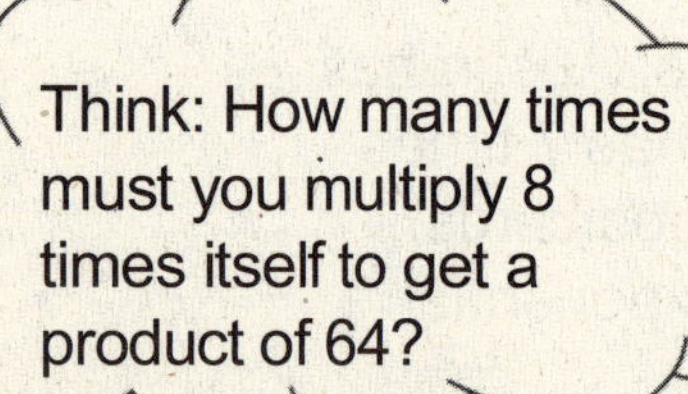

Write each number using an exponent and the given base.

11. 216, base 6: $=$ ______

12. 16, base 4   13. 8, base 2   14. 9, base 3

_______________   _______________   _______________

15. 81, base 9   16. 27, base 3   17. 49, base 7

_______________   _______________   _______________

Holt McDougal Mathematics

Name ____ Date _______________________ Class _______________

Challenge

Money Grows

If an investment of $50 doubles in value, your investment will be
worth $50 \cdot 2 = \$100$.

If it doubles again, your investment will be worth $50 \cdot 2 \cdot 2 = \$200$.
You can also write this as $\$50 \cdot 2^2 = \$200$.

If your money doubles a third time, it will be worth $\$50 \cdot 2^3 = \$400$.

Write the correct answer.

1. If you invest $400, how much will it be worth if it doubles in value twice?

2. If an investment of $550 doubles in value twice, how much will it be worth?

3. If you invest $75, how much will it be worth if it doubles in value 4 times?

4. If you invest $125, how much will it be worth if it doubles in value 4 times?

5. If an investment of $75 triples in value 3 times, how much will it be worth?

6. If an investment of $125 triples in value 4 times, how much will it be worth?

When Jasmine turned 13, she started saving $5 per month until she
turned 18. She then invested her total savings so that it doubled in
value every 7 years.

7. How much money did Jasmine save by the time she turned 18? _______________

8. How many times will her investment double from age 18 to age 60? _______________

9. How much will Jasmine's investment be worth when she turns 60? _______________

When Jake turned 13, he started saving $1.50 per week until he
turned 19. He then invested his total savings so that it tripled in
value every 12 years.

10. How much money did Jake save by the time he turned 19? _______________

11. How many times will his investment triple from age 19 to age 55? _______________

12. How much will Jake's investment be worth when he turns 55? _______________

LESSON 1-2 Problem Solving
Exponents

Write the correct answer.

1. The cells of the bacteria *E. coli* can double every 20 minutes. If you begin with a single cell, how many cells can there be after 4 hours?

2. The population of metropolitan Orlando, Florida, has doubled about every 16 years since 1960. In 2000, the population was 1,644,561. At this doubling rate, what could the population be in 2048?

3. A prizewinner can choose Prize A, $2,000 per year for 15 years, or Prize B, 3 cents the first year, with the amount tripling each year through the fifteenth year. Which prize is more valuable? How much is it worth?

4. Maria had triplets. Each of her 3 children had triplets. If the pattern continued for 2 more generations, how many great-great-grandchildren would Maria have?

Choose the letter for the best answer.

5. A theory states that the CPU clock speed in a computer doubles every 18 months. If the clock speed was 33 MHz in 1991, how can you use exponents to find out how fast the clock speed is after doubling 3 times?

 A 33^3

 B $3^2 \cdot 33$

 C $2^3 \cdot 33$

 D 33^2

6. The classroom is a square with a side length of 13 feet and an area of 169 square feet. How can you write the area in exponential form?

 F 2^{13}

 G 13^2

 H 3^{13}

 J 13^3

7. In 2000, Wake County, North Carolina, had a population of 610,284. This is about twice the population in 1980. If the county grows at the same rate every 20 years, what will its population be in 2040?

 A 915,426

 B 1,220,568

 C 1,830,852

 D 2,441,136

8. The number of cells of a certain type of bacteria doubles every 45 minutes. If you begin with a single cell, how many cells could there be after 6 hours?

 F 64

 G 256

 H 360

 J 540

LESSON 1-2 Reading Strategies
Reading and Understanding Symbols

Exponents are an efficient way to express repeated multiplication.

3^5 ⟶ is read "3 to the fifth power."
3^5 means **3 is a factor 5 times**: $3 \times 3 \times 3 \times 3 \times 3$.
$3^5 = 243$ ⟶ is read "3 to the fifth power equals 243,"
or, "The value of 3 to the fifth power is 243."
The **base** names the factor. The
exponent tells how many times to
repeat the base as a factor.

You can describe some numbers as
the product of a repeated factor.

⟶ $64 = 4 \times 4 \times 4$

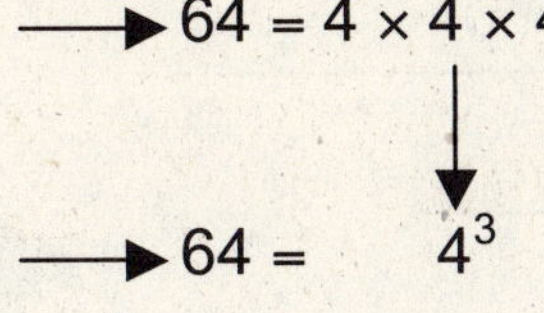

Then write the product with a base
and an exponent.

⟶ $64 = 4^3$

Answer each question.

1. How do you read 5^3? ____

2. What does 5^3 mean? ____

3. What is the value of 5^3? ____

4. Write about the difference between 5^3 and 3^5.

____

____

5. Write 32 using 2 as the repeated factor.

____

6. Write 32 using an exponent and a base of 2. _________________________

7. Write 216 using 6 as the repeated factor.

____

8. Write 216 using an exponent and a base of 6. _________________________

     Holt McDougal Mathematics

LESSON 1-2

Puzzles, Twisters & Teasers

Explore Your Power Base!

Why was the computer tired when it got home?

To find out, solve each problem. Write the letter on the line above the correct answer at the bottom of the page. Each answer must match exactly. Some letters will not be used.

E 3^4 _______________

T   49, base 7 _______________

K 10^3 _______________

B   81, base 9 _______________

I 2^6 _______________

D 5^2 _______________

W 18^1 _______________

C   25, base 5 _______________

A   27, base 3 _______________

H 14^0 _______________

V 12^2 _______________

R   256, base 4 _______________

O   216, base 6 _______________

U 5^3 _______________

S 6^2 _______________

9^2	81	5^2	3^3	125	36	81	64	7^2	1	3^3	25

3^3	1	3^3	4^4	25	25	4^4	64	144	81

    Holt McDougal Mathematics

<table>
<tr><td>LESSON
1-3</td><td></td></tr>
</table>

Practice A
Scientific Notation

Choose the letter for the best answer.

1. $4 \cdot 10^2$

   A   4   C   400

   B   40   D   100

2. $16 \cdot 10^0$

   F   1600   H   10

   G   16   J   0

3. $9 \cdot 10^3$

   A   9,000   C   90

   B   900   D   9

4. $17 \cdot 10^1$

   F   1.7   H   170

   G   17   J   1700

Multiply.

5. $23 \cdot 10^2$

6. $15 \cdot 10^4$

7. $30 \cdot 10^2$

8. $28 \cdot 10^3$

9. $132 \cdot 10^2$

10. $201 \cdot 10^3$

11. $456 \cdot 10^2$

12. $108 \cdot 10^4$

Write the number in scientific notation.

13. 56,000

14. 306,000

15. 8,000,000

16. 7,200,000

17. 14,000,000

18. 41.00

19. 2,144,000

20. $20.3 \cdot 10^5$

21. Lake Huron covers an area of about 23,000 square miles.
Write this number in scientific notation.

22. The planet Mercury is about 3.6×10^7 miles from the sun. Write
this number in standard form.

Holt McDougal Mathematics

LESSON 1-3

Practice B

Scientific Notation

Multiply.

1. $6 \cdot 10^3$

2. $22 \cdot 10^1$

3. $8 \cdot 10^2$

4. $18 \cdot 10^0$

5. $70 \cdot 10^2$

6. $25 \cdot 10^3$

7. $3 \cdot 10^4$

8. $180 \cdot 10^3$

9. $84 \cdot 10^4$

10. $315 \cdot 10^2$

11. $210 \cdot 10^3$

12. $1{,}004 \cdot 10^3$

13. $1{,}764 \cdot 10^1$

14. $856 \cdot 10^0$

15. $4{,}055 \cdot 10^3$

16. $716 \cdot 10^4$

Write each number in scientific notation.

17. 34,000

18. 7,700

19. 2,100,000

20. 404,000

21. 21,000,000

22. 612.00

23. 3,001,000

24. $62.13 \cdot 10^4$

25. Lake Superior covers an area of about 31,700 square miles. Write this number in scientific notation.

26. Mars is about $1.42 \cdot 10^8$ miles from the sun. Write this number in standard form.

27. In 2005, the population of China was about $1.306 \cdot 10^9$. What was the population of China written in standard form?

28. A scientist estimates there are 4,800,000 bacteria in a test tube. How does she record the number using scientific notation?

                                    Holt McDougal Mathematics

LESSON 1-3 **Practice C**
Scientific Notation

Multiply.

1. $5 \cdot 10^3$
2. $471 \cdot 10^2$
3. $39.5 \cdot 10^1$
4. $200 \cdot 10^5$

5. $7{,}025 \cdot 10^0$
6. $5.7 \cdot 10^6$
7. $66.25 \cdot 10^4$
8. $9.01 \cdot 10^9$

Write each number in scientific notation.

9. $25{,}000$
10. $9{,}900$
11. $9{,}700{,}000$
12. $95.6 \cdot 10^8$

13. $23{,}000{,}000{,}000$
14. 110.00
15. $301.9 \cdot 10^5$
16. $73.55 \cdot 10^4$

Write the missing number or numbers.

17. $1.23 \times 10^? = 12{,}300$
18. $8.3 \times 10^5 = ?$
19. $112{,}000{,}000 = ? \times 10^8$

20. $410{,}000 = ? \times 10^5$
21. $7.7 \times 10^7 = ?$
22. $2{,}950{,}000 = 2.95 \times 10^?$

23. The Caspian Sea covers an area of about 143,250 square miles. Write this number in scientific notation.

24. The distance between Jupiter and Saturn is about 4.03×10^8 miles. Write this number in standard form.

25. In 2002, there were about 3.005×10^7 pet dogs in Brazil and about 9.65×10^6 pet dogs in Japan. In which country were there more pet dogs?

26. A scientist estimates there are 37,000,000 bacteria in a petri dish. He records the number using scientific notation. What does he write?

  Holt McDougal Mathematics

Review for Mastery

LESSON 1-3

Scientific Notation

To multiply by a power of 10, use the exponent to find the number of zeros in the product.

- Multiply $42 \cdot 10^4$.

$$42 \cdot 10^4 = 420,000$$

Find each product.

1. $84 \cdot 10^3$

   The product should have ___ zeros.

   $84 \cdot 10^3 =$ _____________

2. $61 \cdot 10^5$

   The product should have ___ zeros.

   $61 \cdot 10^5 =$ _____________

3. $22 \cdot 10^6$

4. $753 \cdot 10^3$

5. $825 \cdot 10^2$

6. $123 \cdot 10^1$

_____________   _____________   _____________   _____________

- Write 926,000 in scientific notation.

First, write the digits before the zeros as a number greater than or equal to 1 and less than 10. The number must have only 1 digit to the left of the decimal point. That digit cannot be zero.

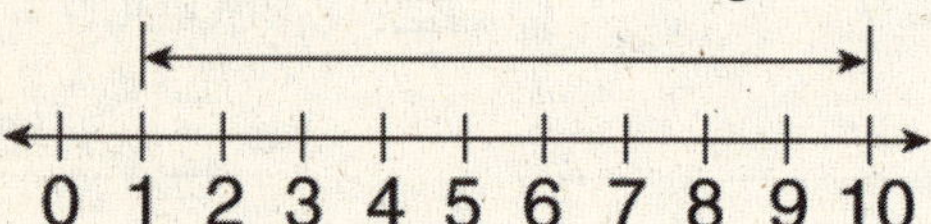

Think: 9.26 is greater than 1 and less than 10.

Then multiply 9.26 by the power of 10 that gives 926,000 as the product.

$$9\,2\,6\,0\,0\,0 = 9.26 \times 10^5$$

Write each number in scientific notation.

7. 5,100

   The decimal point moves _____ places.

   $5,100 =$ _____ . _____ $\times 10$—

8. 1,840,000

   The decimal point moves _____ places.

   $1,840,000 =$ _____ . _____ $\times 10$—

9. 641,000

10. 47,300

11. 8,250,000

12. 703,000

_____________   _____________   _____________   _____________

   Holt McDougal Mathematics

LESSON 1-3

Challenge

Computer Bytes

Each byte in a computer's memory represents about one character. The major units of computer memory are kilobytes (KB), megabytes (MB), and gigabytes (GB).

1 kilobyte = 1,000 bytes	1 KB = 1,000 bytes
1 megabyte = 1,000 kilobytes	1 MB = 1,000 KB
1 gigabyte = 1,000 megabytes	1 GB = 1,000 MB

Write your answers using scientific notation.

1. In 1984, many personal computers had 64 KB of active (RAM) memory. How many bytes does this represent?

2. In 1992, many personal computers had 40 MB of hard drive memory. How many bytes does this represent?

3. In 1997, many personal computers had 1 GB of hard drive memory. How many bytes does this represent?

4. By 2005, many personal computers had 250 GB of hard drive memory. How many bytes does this represent?

Ming saved his computer files on floppy disks. Each disk holds up to 1.44 MB of memory. He used these disks to transfer his files to another computer.

5. How many bytes could each floppy disk hold?

6. Ming's new computer has 120 GB of memory. How many disks could he transfer if each disk held 1.2 MB?

Rachel decided to back up her hard drive's computer files by copying them onto compact disks (CDs). Each CD can hold up to 650 MB of memory, but Rachel saves only 600 MB on each.

7. How many bytes could each CD potentially hold?

8. If Rachel backs up 6 GB of memory, how many bytes of memory will she need?

Holt McDougal Mathematics

LESSON	**Problem Solving**
1-3	*Scientific Notation*

Write the correct answer.

1. Earth is about 150,000,000 kilometers from the sun. Write this distance in scientific notation.

2. The planet Neptune is about 4.5×10^9 kilometers from the sun. Write this distance in standard form.

3. At the end of 2004, the U.S. federal debt was about $7 trillion, 600 billion. Write the amount of the debt in standard form and in scientific notation.

4. Canada is about 1.0×10^7 square kilometers in size. Brazil is about 8,500,000 square kilometers in size. Which country has a greater area?

Choose the letter for the best answer.

5. China's population in 2001 was approximately 1,273,000,000. Mexico's population for the same year was about 1.02×10^8. How much greater was China's population than Mexico's?

   A 1,375,000,000

   B 1,274,020,000

   C 1,171,000,000

   D 102,000,000

6. In mid-2001, the world population was approximately 6.137×10^9. By 2050, the population is projected to be 9.036×10^9. By how much will world population increase?

   F 151,730,000

   G 289,900,000

   H 1,517,300,000

   J 2,899,000,000

7. The Alpha Centauri star system is about 4.3 light-years from Earth. One light-year, the distance light travels in 1 year, is about 6 trillion miles. About how many miles away from Earth is Alpha Centauri?

   A 2.58×10^{13} miles

   B 6×10^{13} miles

   C 1.03×10^{12} miles

   D 2.58×10^9 miles

8. In the fall of 2001, students in Columbia, South Carolina, raised $440,000 to buy a new fire truck for New York City. If the money had been collected in pennies, how many pennies would that have been?

   F 4.4×10^6

   G 4.4×10^5

   H 4.4×10^7

   J 4.4×10^8

LESSON 1-3	# Reading Strategies
	Follow a Procedure

When you have an exponent with a base of 10, the number is called a **power of 10.**

You can use some simple rules to find products with powers of 10.

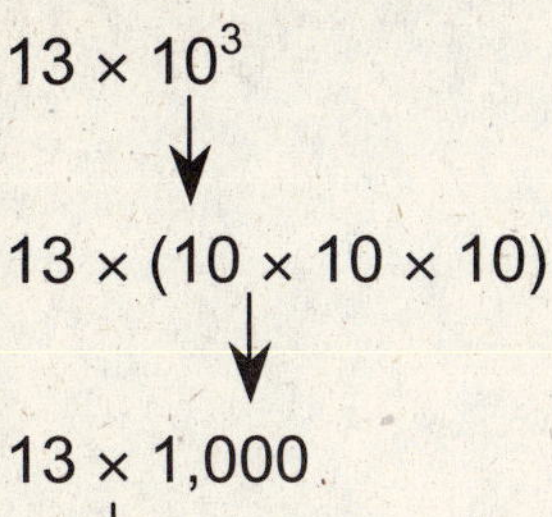

13×10^3

↓

$13 \times (10 \times 10 \times 10)$

↓

$13 \times 1,000$

↓

13.0.0.0.

Move the decimal point
3 places to the right.
You need to add 3 zeros.

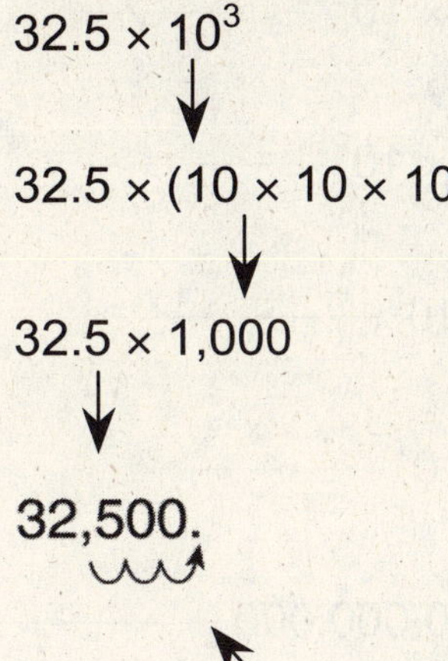

32.5×10^3

↓

$32.5 \times (10 \times 10 \times 10)$

↓

$32.5 \times 1,000$

↓

32,500.

Move the decimal point
3 places to the right.
You need to add 2 zeros.

You can use powers of 10 to write large numbers in **scientific notation.** Scientific notation is used as a shortcut to write very large or very small numbers.

To write 268,000,000 in scientific notation:

Step 1: Move the decimal point to create a number between 1 and 10.   2.6.8.0.0.0.0.0.0. ← Move the decimal point 8 places.

↓

Step 2: The number of places the decimal point is moved is the value of the exponent. 2.68×10^8 ← So, the exponent is 8.

268,000,000 written in scientific notation is 2.68×10^8.

Use 2.8×10^5 to answer Exercises 1–4.

1. How many times is 10 a factor? ____

2. Rewrite the number with 10 as a repeated factor.

____

3. How many places will you move the decimal point?
   How many zeros will be in the product? ___________________________________

4. What is the product of 2.8×10^5? _____________________________________

 Holt McDougal Mathematics

LESSON 1-3 Puzzles, Twisters & Teasers

Oh, the Power of Tens!

Substitute the correct number for the letter or letters in each equation. Use your answers to solve the riddle.

1. $24{,}500 = 2.45 \times 10^E$ ______

2. $280{,}000 = 2.8 \times 10^P$ ______

3. $592{,}000 = I \times 10^5$ ______

4. $16{,}800 = C \times 10^4$ ______

5. $5.4 \times 10^H = 540{,}000{,}000$ ______

6. $24{,}400{,}000 = S \times 10^A$ ______

What's a Martian's favorite snack?

2.44	5	7	1.68	4		1.68	8	5.92	5	2.44

 Holt McDougal Mathematics

LESSON 1-4

Practice A

Order of Operations

Choose the letter for the best answer.

1. $75 + 12 \cdot 2$

   A 87   C 108

   B 99   D 174

2. $100 - 25 \div 5$

   F 15   H 80

   G 75   J 95

3. $50 - 18 \div 6 + 2$

   A 49   C 10

   B 40   D 4

4. $72 - 4^2 \cdot 2$

   F 32   H 56

   G 40   J 64

5. $(8 + 22) \div 5 + 5$

   A 30   C 11

   B 17.4   D 3

6. $3^3 - (9 \cdot 2 + 1)$

   F 19   H 8

   G 10   J −10

Simplify each expression.

7. $2^4 \div 8 + 5$

8. $18 + 2(1 + 3^2)$

9. $(16 \div 4) + 4 \cdot (2^2 - 2)$

10. $2^3 - (3 \cdot 5 - 8)$

11. $35 + 4^2 - (6 - 3)$

12. $6 \cdot 7 - 3(4 + 1)$

13. $100 \div 5 \cdot 2^2$

14. $(5 + 2)^2 \div 7 - 6$

15. $15 - 3 \cdot 4 \div 2 + 5$

16. Leon rents a video game for $5. He returns the video game 3 days late. The late fee is $1 for each day late. Simplify the expression $5 + 3 \cdot 1$ to find out how much it costs Leon to rent the video game.

17. Olivia shovels snow for 6 neighbors. Each neighbor pays her $8. Two neighbors each give her a $2 tip as well. Simplify the expression $6 \cdot 8 + 2 \cdot 2$ to find out how much money Olivia earns in all.

   Holt McDougal Mathematics

LESSON 1-4 · Practice B
Order of Operations

Simplify each expression.

1. $15 \cdot 3 + 12 \cdot 2$

2. $212 + 21 \div 3$

3. $9 \cdot 3 - 18 \div 3$

4. $65 - 36 \div 3$

5. $100 - 9^2 + 2$

6. $3 \cdot 5 - 45 \div 3^2$

7. $54 \div 6 + 4 \cdot 6$

8. $(6 + 5) \cdot 16 \div 2$

9. $60 - 8 \cdot 12 \div 3$

10. $45 - 3^2 \cdot 5$

11. $52 - (8 \cdot 2 \div 4) + 3^2$

12. $(2^3 + 10 \div 2) \cdot 3$

13. $25 + 7(18 - 4^2)$

14. $(6 \cdot 3 - 12)^2 \div 9 + 7$

15. $4^3 - (3 + 12 \cdot 2 - 9)$

16. $2^4 \div 8 + 5$

17. $(1 + 2)^2 \cdot (3 - 1)^2 \div 2$

18. $(16 \div 4) + 4 \cdot (2^2 - 2)$

19. $2^5 - (3 \cdot 7 - 7)$

20. $75 + 5^2 - (8 - 3)$

21. $9 \cdot 6 - 5(10 - 3)$

22. $96 \div 4 + 5 \cdot 2^2$

23. $(15 - 6)^2 \div 3 - 3^3$

24. $19 - 8 \cdot 5 \div 10 + 6 \div 3$

25. Jared has $32. He buys 5 packs of trading cards that cost $3 each and a display book that costs $7. Simplify the expression $32 - (5 \cdot 3 + 7)$ to find out how much money Jared has left.

26. David buys 3 movie tickets for $6 each and 2 bags of popcorn for $2 each. Simplify the expression $3 \cdot 6 + 2 \cdot 2$ to find out how much money David spent in all.

Holt McDougal Mathematics

LESSON 1-4

Practice C
Order of Operations

Simplify each expression.

1. $25 \cdot 3 + 60 \cdot 2$

2. $350 \div 5 + 12 \cdot 7$

3. $3 \cdot 9 + 96 \div 4$

4. $77 - 42 \div 7^1$

5. $532 - 2^5 \div 4$

6. $3(20 - 4^2) + 7$

7. $270 \div 6 + 6^2$

8. $(5 + 6)^2 + 18 \div 2$

9. $10^2 - 25 \cdot 3 \div 5$

10. $65 - 4^3 \cdot 1^7$

11. $40 - (5 \cdot 2) + 8$

12. $(6^2 + 4) \div 5$

13. $2^4 \div 8 + 5$

14. $(1 + 2)^2 \cdot (3 - 1)^2 \div 2$

15. $(16 \div 4) + 4 \cdot (2^2 - 2)$

Insert grouping symbols to make each statement true.

16. $18 + 2 \cdot 1 + 3^2 = 38$

17. $4 \cdot 2 - 2^2 \div 9 + 2 = 6$

18. $3^3 - 9 \cdot 2 + 1 = 8$

19. $2^3 - 3 \cdot 5 - 8 = 1$

20. $35 + 4^2 - 6 - 3 = 48$

21. $6 \cdot 7 - 3 \cdot 4 + 1 = 27$

22. A group of students charges \$7 to clean the exterior and \$6 to clean the interior of a car. They clean 9 exteriors and 5 interiors. Simplify the expression $7 \cdot 9 + 6 \cdot 5$ to find out how much money the students raised in all.

23. Ariel has \$65. She buys 5 books that cost \$8.00 each, a bookmark that costs \$2.00, and a magazine that costs \$4.00. Simplify the expression $65 - (5 \cdot 8 + 2 + 4)$ to find out how much money Ariel has left.

Holt McDougal Mathematics

LESSON 1-4 — Review for Mastery
Order of Operations

To help you remember the order of operations use the phrase
"**P**lease **E**xcuse **M**y **D**ear **A**unt **S**ally."

> **P**: first, **p**arentheses (if any)
> **E**: second, **e**xponents (if any)
> **M** and **D**: then, **m**ultiplication and **d**ivision, in order from left to right
> **A** and **S**: finally, **a**ddition and **s**ubtraction, in order from left to right

Evaluate.

$$39 \div (9 + 4) + 5 - 2^2$$

Parentheses $\longrightarrow$ $39 \div 13 + 5 - 2^2$

Exponents $\longrightarrow$ $39 \div 13 + 5 - 4$

Multiply and divide from left to right $\longrightarrow$ $3 + 5 - 4$

Add and subtract from left to right $\longrightarrow$ $8 - 4 = 4$

Simplify each expression.

1. $12 \cdot 4 - 2$

   ____ $- 2$

   ____

2. $15 \div 3 \cdot 5$

   ____ $\cdot 5$

   ____

3. $15 \cdot 3 \div 5$

   ____ $\div 5$

   ____

4. $8 + 20 \div 4$

   ____

5. $5 - 2 \cdot 6 \div 4 + 1$

   ____

6. $3^2 + 6 \cdot 4 - 5^2$

   ____

7. $1 + 4 \cdot 9 \div 6 - 7$

   ____

8. $18 \div (6 \div 3)$

   ____

9. $(18 \div 6) \div 3$

   ____

10. $4 \cdot 5 + 8 \div 2 - 7$

    ____

11. $2 \cdot 3 - 8 \div 2^2$

    ____

12. $8(7 - 6) \div 2^3$

    ____

 Holt McDougal Mathematics

Challenge
LESSON 1-4

Fixed and Variable Costs

A *fixed cost* is a one-time cost. A *variable cost* changes depending on your use of a product or a service.

The annual enrollment fee per year at a fitness club (fixed cost) is $30. You also pay $2 per visit (variable cost), and you visit the club 8 times per month. What is your total annual cost?

$2 \cdot (8 \cdot 12)$ variable cost times total visits

$30 + 2 \cdot (8 \cdot 12)$ total annual cost

222 Your total annual cost is $222.

Use the information above to solve problems 1–4.

1. Suppose the annual fee is $25, but the cost per visit is $3. What is your annual cost?

2. Suppose you visit the club 3 times per week instead of 8 times per month. What is your annual cost?

3. Suppose the cost per visit is $3 after the first 50 visits per year. What is your annual cost?

4. Suppose you pay for up to 75 visits per year. Any additional visits are free. What is your annual cost?

The school band is raising money for a trip. The members ordered 5 dozen jerseys for $7 each and sold them for $12 each. They also ordered 4 dozen sweatshirts for $11 each and sold them for $18 each. The band paid $35 to create the design.

5. Write and simplify an expression to calculate the band's variable costs for the clothing.

6. Write and simplify an expression to calculate the band's total cost, including fixed costs.

7. Write and simplify an expression to calculate the band's profit.

8. What would the profit be if jerseys sold for $10 and sweatshirts for $20?

      Holt McDougal Mathematics

LESSON 1-4

Problem Solving

Order of Operations

Write the correct answer.

1. In 1975, the minimum wage was $2.10 per hour. Write and simplify an expression to show wages earned in a 35-hour week after a $12 tax deduction.

2. George bought 3 boxes of Girl Scout cookies at $3.50 per box and 4 boxes at $3.00 per box. Write and simplify an expression to show his total cost.

3. In 1 week Ed works 4 days, 3 hours a day, for $12 per hour, and 2 days, 6 hours a day, for $15 per hour. Simplify the expression $12(4 \cdot 3) + 15(2 \cdot 6)$ to find Ed's weekly earnings.

4. Keisha had $150. She bought jeans for $27, a sweater for $32, 3 blouses for $16 each, and 2 pairs of socks for $6 each. Simplify the expression $150 - [27 + 32 + (3 \cdot 16) + (2 \cdot 6)]$ to find out how much money she has left.

Choose the letter for the best answer.

5. As of September 1, 1997, the minimum wage was set at $5.15 per hour. How much more would someone earn now than in 1997 if she earns $5 more per hour for a 40-hour week?

   A $206 more

   B $200 more

   C $406 more

   D $400 more

6. Gary received $200 in birthday gifts. He bought 5 CDs for $15 each, 2 posters for $12 each, and a $70 jacket. How much money does he have left?

   F $31

   G $10

   H $132

   J $169

7. Yvonne took her younger brother and his friends to the movies. She bought 5 tickets for $8 each, 4 drinks for $2 each, and two $3 containers of popcorn. How much did she spend?

   A $22

   B $51

   C $54

   D $38

8. On a business trip, Mr. Chang stayed in a hotel for 7 nights. He paid $149 per night. While he was there, he made 8 phone calls at $2 each and charged $81 to room service. How much did he spend?

   F $246

   G $946

   H $1,043

   J $1,140

   Holt McDougal Mathematics

LESSON 1-4 Reading Strategies
Use a Flowchart

When you read a book, you read from left to right. When you evaluate an expression, you cannot always work from left to right. You must follow a special rule called the **order of operations**. Use the flowchart below to help you follow the order of operations.

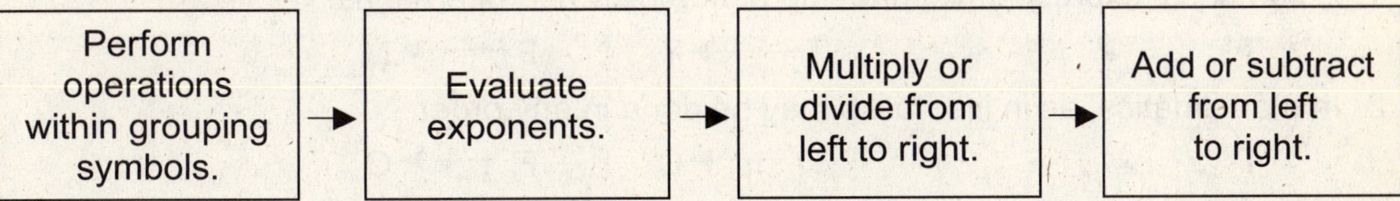

Example

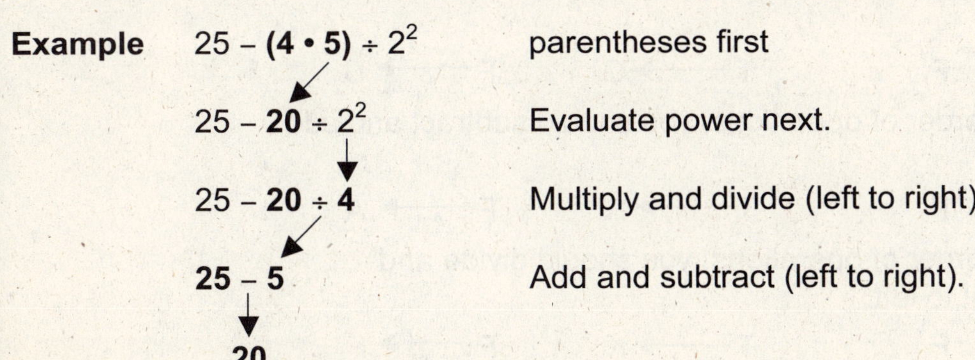

$25 - (4 \cdot 5) \div 2^2$	parentheses first
$25 - 20 \div 2^2$	Evaluate power next.
$25 - 20 \div 4$	Multiply and divide (left to right).
$25 - 5$	Add and subtract (left to right).
20	

Answer each question.

1. In what order will you perform the operations in the following expression: $30 - 18 \div 2 \cdot 3 + 4$?

   ____

2. Simplify this expression: $30 - 18 \div 2 \cdot 3 + 4$. _______________________

3. Make a flow chart for the order of operations in this expression:
   $3^3 - (2 + 5 \cdot 2)$.

   ____

4. Simplify this expression: $3^3 - (2 + 5 \cdot 2)$. _______________________

5. Make a flow chart for the order of operations in this expression:
   $(3 + 12 \div 3)^2 - 4$.

   ____

6. Simplify this expression: $(3 + 12 \div 3)^2 - 4$. _______________________

          Holt McDougal Mathematics

LESSON 1-4 — Puzzles, Twisters & Teasers

Is Everything in Order?

What's the last thing you take off before you go to bed?

Decide whether each statement below is true or false. Use your answers to solve the riddle.

1. A numerical expression is made up of numbers and operations.

   T   F   T ——▶ Y   F ——▶ M

2. In mathematics, as in life, tasks may be done in any order.

   T   F   T ——▶ C   F ——▶ O

3. When using the order of operations, you should do division after subtraction.

   T   F   T ——▶ G   F ——▶ U

4. When using the order of operations, you should subtract and add from left to right.

   T   F   T ——▶ R   F ——▶ A

5. When using the order of operations, you should divide and multiply from right to left.

   T   F   T ——▶ K   F ——▶ F

6. When an expression has a set of grouping symbols within a second set of grouping symbols, you should begin with the innermost set.

   T   F   T ——▶ E   F ——▶ D

7. You should perform operations inside parentheses first.

   T   F   T ——▶ T   F ——▶ I

8. When using the order of operations, you should evaluate the power expression after multiplying and adding.

   T   F   T ——▶ B   F ——▶ H

9. Mathematicians agree on using the order of operations.

   T   F   T ——▶ L   F ——▶ J

You take ___ ___ ___ ___   ___ ___ ___ ___   ___ ___ ___
     1   2   3   4   5   6   6   7   2   5   5

___ ___ ___   ___ ___ ___ ___ ___
 7   8   6   5   9   2   2   4

Holt McDougal Mathematics

LESSON 1-5 · Practice A
Properties of Numbers

Tell which property is shown.

1. $5 + 0 = 5$

____

2. $8 \cdot (6 \cdot 2) = (8 \cdot 6) \cdot 2$

____

3. $9 + 8 = 8 + 9$

____

4. $4 \cdot 1 = 4$

____

Simplify each expression. Write a reason for each step.

5. $13 + 28 + 7$

$13 + 28 + 7 = 28 + 13 + 7$ Reason: Commutative Property

$= 28 + (13 + 7)$ Reason: _______________

$= 28 + \underline{\hspace{2em}}$ Reason: Add.

$= \underline{\hspace{2em}}$ Reason: _______________

6. $20 \cdot (17 \cdot 5)$

$20 \cdot (17 \cdot 5) = 20 \cdot (\underline{\hspace{2em}} \cdot 17)$ Reason: _______________

$= (20 \cdot \underline{\hspace{2em}}) \cdot 17$ Reason: _______________

$= \underline{\hspace{3em}} \cdot \underline{\hspace{2em}}$ Reason: Multiply.

$= \underline{\hspace{3em}}$ Reason: _______________

Use the Distributive Property to find each product.

7. $4(17)$

$4(17) = 4 \cdot (10 + \underline{\hspace{2em}})$

$= (4 \cdot \underline{\hspace{2em}}) + (4 \cdot 7)$

$= \underline{\hspace{2em}} + \underline{\hspace{2em}}$

$= \underline{\hspace{2em}}$

8. $3(28)$

$3(28) = \underline{\hspace{4em}}$

$= \underline{\hspace{4em}}$

$= \underline{\hspace{4em}}$

$= \underline{\hspace{4em}}$

     Holt McDougal Mathematics

LESSON 1-5

Practice B
Properties of Numbers

Tell which property is represented.

1. $12 \cdot 14 = 14 \cdot 12$

2. $1 \cdot 36 = 36$

____

3. $(17 + 36) + 4 = 17 + (36 + 4)$

4. $8 \cdot 12 \cdot 5 = 8 \cdot (12 \cdot 5)$

____

Simplify each expression. Justify each step.

5. $4 \cdot 9 \cdot 50$

$4 \cdot 9 \cdot 50 = $ _______________________________

$= $ _______________________________

$= $ _______________________________

$= $ _______________________________

6. $(33 + 45) + 7$

$(33 + 45) + 7 = $ _______________________________

$= $ _______________________________

$= $ _______________________________

$= $ _______________________________

Use the Distributive Property to find each product.

7. $3(26) = $ _______________________

8. $(18)9 = $ _______________________

$= $ _______________________

$= $ _______________________

$= $ _______________________

$= $ _______________________

LESSON 1-5 Practice C
Properties of Numbers

Complete each equation. Then tell which property is represented.

1. $23 + \underline{\hspace{1cm}} = 23$

2. $\underline{\hspace{1cm}} \cdot (19 + 6) = (7 \cdot 19) + (7 \cdot 6)$

3. $27 + 45 = \underline{\hspace{1cm}} + 27$

4. $6 \cdot (\underline{\hspace{1cm}} \cdot 7) = (6 \cdot 14) \cdot 7$

Simplify each expression. Justify each step.

5. $(40 \cdot 7) \cdot 5$

$(40 \cdot 7) \cdot 5 = \underline{\hspace{4cm}}$

$= \underline{\hspace{4cm}}$

$= \underline{\hspace{4cm}}$

$= \underline{\hspace{4cm}}$

6. $15 + 98 + 85$

$15 + 98 + 85 = \underline{\hspace{4cm}}$

$= \underline{\hspace{4cm}}$

$= \underline{\hspace{4cm}}$

$= \underline{\hspace{4cm}}$

Use the Distributive Property to find each product.

7. $7(43) = \underline{\hspace{3cm}}$

$= \underline{\hspace{3cm}}$

$= \underline{\hspace{3cm}}$

8. $(597)4 = \underline{\hspace{3cm}}$

$= \underline{\hspace{3cm}}$

$= \underline{\hspace{3cm}}$

  Holt McDougal Mathematics

LESSON 1-5

Review for Mastery

Properties of Numbers

You can use the Commutative Property, the Associative Property, and the Distributive Property with mental math to simplify expressions.

$16 + 47 + 14 = 47 + 16 + 14$	Commutative Property	$8 \cdot 3 \cdot 5 = 3 \cdot 8 \cdot 5$
$= 47 + (16 + 14)$	Associative Property	$= 3 \cdot (8 \cdot 5)$
$= 47 + 30$	Mental math	$= 3 \cdot 40$
$= 77$	Mental math	$= 120$
$9(28) = 9(20 + 8)$		$9(28) = 9(30 - 2)$
$= (9 \cdot 20) + (9 \cdot 8)$	Distributive Property	$= (9 \cdot 30) - (9 \cdot 2)$
$= 180 + 72$	Mental math	$= 270 - 18$
$= 252$	Mental math	$= 252$

Simplify each expression. Tell what properties you used.

1. $(45 + 39) + 25 = (39 + \underline{\quad}) + 25$ _____________________ Property

$= 39 + (\underline{\quad} + \underline{\quad})$ _____________________ Property

$= 39 + \underline{\quad}$

$= \underline{\quad}$

2. $25 \cdot 7 \cdot 4 = 25 \cdot \underline{\quad} \cdot \underline{\quad}$ _____________________ Property

$= (\underline{\quad} \cdot \underline{\quad}) \cdot \underline{\quad}$ _____________________ Property

$= \underline{\quad} \cdot \underline{\quad}$

$= \underline{\quad}$

3. $5(18) = 5 \cdot (10 + \underline{\quad})$

$= (5 \cdot \underline{\quad}) + (5 \cdot \underline{\quad})$

$= \underline{\quad} + \underline{\quad}$

$= \underline{\quad}$

_____________________ Property

4. $6(29) = 6 \cdot (30 - \underline{\quad})$

$= (6 \cdot \underline{\quad}) - (6 \cdot \underline{\quad})$

$= \underline{\quad} - \underline{\quad}$

$= \underline{\quad}$

_____________________ Property

      Holt McDougal Mathematics

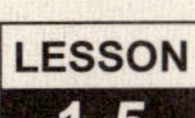

Challenge

LESSON 1-5

What's the Expression?

On each tic-tac-toe board below, exactly one row, column, or
diagonal contains three expressions with the same value.

**Predict which row, column, or diagonal of squares will have
three expressions with the same value. Draw a line through
those three squares. Then check your prediction. Find the value
of each expression on the board.**

$9(80 - 12)$	$1 \cdot 621$	$(2 + 9)34$
____	____	____
$9(30 + 4)$	$2 \cdot 9 \cdot 34$	$(3 \cdot 3)60$
____	____	____
$2(9 + 34)$	$9(70 - 2)$	$9(60 + 8)$
____	____	____

$4 \cdot 3)(8 \cdot 8)$	$12(80 - 8)$	$12(60 + 12)$
____	____	____
$12(60 + 4)$	$(2 \cdot 2)64$	$1 \cdot (12 \cdot 72)$
____	____	____
$(70 - 6)4$	$4 \cdot (3 + 72)$	$(4 \cdot 3)72$
____	____	____

Holt McDougal Mathematics

LESSON 1-5

Problem Solving
Properties of Numbers

Write the correct answer.

1. Jo makes and sells jewelry. She sold three bracelets for $45, $17, and $25. Write an expression for the total Jo received. Explain how you can use properties and mental math to simplify the expression.

2. Use parentheses to show two ways of grouping the numbers in $12 \cdot 8 \cdot 25$. Tell which expression you think would be easier to simplify, and why. Then simplify the expression.

3. The distance from Mark's apartment to his job is 27 miles. Mark works 5 days per week. How many miles does Mark drive to and from work each week?

4. Jane said that $6(64) = 6(50) + 6(14)$. Is she correct? Use the Distributive Property to explain your answer.

Choose the letter for the best answer.

5. Maxine works 8 hours at a rate of $16 per hour. Which expression could **not** be used to find her total earnings in dollars?

   A $8 \cdot (10 \cdot 6)$

   B $8 \cdot (20 - 4)$

   C $8 \cdot (10 + 6)$

   D $8 \cdot (8 + 8)$

6. Rosemary runs 16 miles on Friday, 8 miles on Saturday, and 14 miles on Sunday. How many miles does Rosemary run in all?

   F 22 mi

   G 24 mi

   H 30 mi

   J 38 mi

7. Which of the following represents the Identity Property?

   A $(8 \cdot 4) \cdot 3 = 8 \cdot (4 \cdot 3)$

   B $16 \cdot 0 = 0$

   C $25 \cdot 1 = 25$

   D $6(26) = 6(20) + 6(6)$

8. Which of the following shows how the Distributive Property could be used to simplify $7(28)$?

   F $7 \cdot 2 \cdot 8$

   G $7 \cdot (20 \cdot 8)$

   H $7 \cdot (20 + 8)$

   J $(7 \cdot 20) + 8$

<table><tr><td>**LESSON**
1-5</td><td>

Reading Strategies
Use a Flowchart
</td></tr></table>

Use a flowchart to help you simplify an expression, such as $(25 + 89) + 15$.

Step 1: Choose two numbers that are easy to add.
$(25 + 89) + 15$

↓

Step 2: Rewrite the expression so the two numbers are next to each other.
Use the Commutative Property. $(25 + 89) + 15 = (89 + 25) + 15)$

↓

Step 3: Rewrite the expression so the two numbers are grouped together.
Use the Associative Property. $(89 + 25) + 15 = 89 + (25 + 15)$

↓

Step 4: Add.
$89 + (25 + 15) = 89 + 40 = 129$

Use the expression $16 + (39 + 14)$ for Exercises 1–4.

1. Which two numbers are easy to add? ___________________________

2. Rewrite the expression so that the numbers that are easy
   to add are next to each other. What property lets you do this?

   ____

3. Rewrite the expression so that the numbers that are easy to add
   are grouped together. What property lets you do this?

   ____

4. Simplify the expression. ___________________________________

Use the expression $35 + 47 + 5$ for Exercises 5–8.

5. Which two numbers are easy to add? ___________________________

6. Rewrite the expression so that the numbers that are easy to add
   are next to each other. What property lets you do this?

   ____

7. Rewrite the expression so that the numbers that are easy to add
   are grouped together. What property lets you do this?

   ____

8. Simplify the expression. ___________________________________

LESSON 1-5 Puzzles, Twisters, & Teasers
Make the Connection!

Draw a line to connect each equation to the property it represents.

1. $3 \cdot (8 \cdot 5) = (3 \cdot 8) \cdot 5$ Identity Property   **R**

2. $1 \cdot 15 = 15$ Commutative Property   **Y**

3. $9(4 + 6) = 9(4) + 9(6)$ Associative Property   **E**

4. $38 + 7 = 7 + 38$ Distributive Property   **P**

Draw a line to connect expressions with the same value.

5. $5 + (7 + 6)$ $5(7) + 5(6)$ **S**

6. $5(70 + 6)$ $5(70) - 5(6)$ **H**

7. $5 + 70 + 6$ $(5 + 7) + 6$ **T**

8. $5 \cdot 7 \cdot 6$ $(5 \cdot 70) \cdot 6$ **I**

9. $5(13)$ $7 \cdot (5 \cdot 6)$ **D**

10. $5(70 - 6)$ $70 + 11$ **L**

11. $5 \cdot (70 \cdot 6)$ $5(70) + 5(6)$ **K**

Start with number 1. Find the letter next to the answer. Use the letters to find out why potatoes are good detectives.

$$\overline{}_{5} \ \overline{}_{10} \ \overline{}_{1} \ \overline{}_{4} \qquad \overline{}_{6} \ \overline{}_{1} \ \overline{}_{1} \ \overline{}_{3}$$

$$\overline{}_{5} \ \overline{}_{10} \ \overline{}_{1} \ \overline{}_{11} \ \overline{}_{2} \qquad \overline{}_{1} \ \overline{}_{4} \ \overline{}_{1} \ \overline{}_{9}$$

$$\overline{}_{3} \ \overline{}_{1} \ \overline{}_{1} \ \overline{}_{7} \ \overline{}_{1} \ \overline{}_{8}$$

     Holt McDougal Mathematics

Dear Family,

In this section, the student will learn the basic building blocks of algebra, beginning with recognizing the difference between a **constant** and a **variable.** The student will learn to **evaluate** an **algebraic expression** by substituting a given value for a variable and following the order of operations. The student will follow the method below no matter how many variables are given in the expression.

Evaluate $5b + 7$ for $b = 6$.

$5b + 7$
$5(6) + 7$ Substitute 6 for b.
$30 + 7$ Multiply before adding.
37 Add.

Being able to recognize math terminology is an important skill when solving word problems. Listed below are some key phrases that may help the student translate math words into algebraic expressions.

Operation	Key Word Phrases	Algebraic Expression
Addition +	a number plus 7 add 7 to a number the sum of a number and 7 7 more than a number	$m + 7$
Subtraction −	a number minus 10 10 less than a number a number decreased by 10 the difference of a number and 10	$m - 10$
Multiplication ×	9 times a number the product of 9 and a number	$9m$
Division ÷	a number divided by 12 12 divided into a number the quotient of a number and 12	$m \div 12$ or $\dfrac{m}{12}$

Vocabulary

These are the math words we are learning:

Addition Property of Equality You can add the same amount to both sides of an equation, and the statement will still be true.

algebraic expression an expression that consists of one or more variables, and may contain constants and operations

coefficient the numerical factor in a term containing a variable

constant a number that does not change in an expression

Division Property of Equality You can divide both sides of an equation by the same nonzero number, and the statement will still be true.

equation a mathematical statement that says two expressions are equal in value

evaluate substitute a number for a variable in an algebraic expression

inverse operations operations that undo each other

   Holt McDougal Mathematics

Family Letter

1B Algebraic Thinking *continued*

As a lead-in to solving equations, the student will learn to combine like terms. *Like terms* are terms with the same variables raised to the same exponents. Combining like terms is similar to grouping objects that are alike.

The student will determine whether a solution makes an equation true and learn how to solve one-step equations. He or she will use inverse operations to isolate a variable in an equation. This is how the student will solve one-step equations.

<table>
<tr><td>

Solve an equation using addition.

Solve the equation.

$x - 24 = 65$

Think: 24 is subtracted from x, so add 24 to both sides to isolate x.

$$\begin{array}{rcl} x - 24 &=& 65 \\ +24 & & +24 \\ \hline x &=& 89 \end{array}$$

</td><td>

Solve an equation using subtraction.

Solve the equation.

$y + 18 = 61$

Think: 18 is added to y, so subtract 18 from both sides to isolate y.

$$\begin{array}{rcl} y + 18 &=& 61 \\ -18 & & -18 \\ \hline y &=& 43 \end{array}$$

</td></tr>
<tr><td>

Solve an equation using multiplication.

Solve the equation.

$$\frac{k}{5} = 60$$

Think: k is divided by 5, so multiply both sides by 5 to isolate k.

$$(5)\frac{k}{5} = 60(5)$$
$$k = 300$$

</td><td>

Solve an equation using division.

Solve the equation.

$180 = 6s$

Think: s is multiplied by 6, so divide both sides by 6 to isolate s.

$$\frac{180}{6} = \frac{6s}{6}$$
$$30 = s$$

</td></tr>
</table>

Encourage the student to explain how to solve different equations in order to strengthen his or her newly learned skills.

Sincerely,

> **Multiplication Property of Equality** You can multiply both sides of an equation by the same number, and the statement will still be true.
>
> **solution** a value of the variable that makes an equation true
>
> **Subtraction Property of Equality** You can subtract the same amount from both sides of an equation, and the statement will still be true.
>
> **term** a number, variable, or a product of numbers and variables
>
> **variable** a letter that represents a number that can vary in value

Holt McDougal Mathematics

Name _________________________ Date _____________ Class _____________

At-Home Practice

1B Algebraic Thinking

Evaluate $n + 6$ for each value of n.

1. $n = 5$ 2. $n = 15$ 3. $n = 12$ 4. $n = 27$

_________   _________   _________   _________

Evaluate each expression for the given value of the variable.

5. $8y + 3$ for $y = 4$ 6. $11w - 28$ for $w = 7$ 7. $3t^2 + 6$ for $t = 4$

_________   _________   _________

Write each phrase as an algebraic expression.

8. the difference of a number and 11   9. the product of 5 and a number

_________   _________

10. 13 less than 3 times a number   11. 8 more than 5 times a number

_________   _________

Identify like terms.

12. $13a, 3a^3, 3a, x^3, 2x, 5a$ 13. $r^2, 2r, 5s^2, 5, 2r^2, rs$

_________   _________

Simplify.

14. $7x + 5x$ 15. $2s + 5s + 2s$ 16. $14 + 4f + 3f^3 + 8f + 7$

_________   _________   _________

Determine whether the given value of the variable is a solution.

17. $x = 2$ for $5x + 3 = 13$ 18. $t = 4$ for $5t - 8 + 2t = 30$

_________   _________

Solve the equation. Check your answer.

19. $c + 61 = 105$ 20. $4v = 328$ 21. $\dfrac{w}{6} = 87$ 22. $t - 97 = 54$

_________   _________   _________   _________

Answers: 1. 11 **2.** 21 **3.** 18 **4.** 33 **5.** 35 **6.** 49 **7.** 54 **8.** $n - 11$ **9.** $5n$ **10.** $3n - 13$ **11.** $5n + 8$ **12.** $13a, 3a, 5a$ **13.** $r^2, 2r^2$ **14.** $12x$ **15.** $9s$ **16.** $3f^3 + 12f + 21$ **17.** yes **18.** no **19.** $c = 44$ **20.** $v = 82$ **21.** $w = 522$ **22.** $t = 151$

     Holt McDougal Mathematics

CHAPTER 1

Family Fun
Sea-Mail

Solve each equation. Then put the letter of the variable above its value to answer the riddle.

$14d = 84$ $d =$ _______________

$32 = y + 21$ $y =$ _______________

$\dfrac{o}{4} = 6$ $o =$ _______________

$r - 3 = 15$ $r =$ _______________

$25c = 125$ $c =$ _______________

$s + 17 = 26$ $s =$ _______________

$\dfrac{n}{5} = 8$ $n =$ _______________

$b - 7 = 5$ $b =$ _______________

$180 = 12e$ $e =$ _______________

How did the Vikings send secret messages?

___ ___   ___ ___ ___ ___ ___   ___ ___ ___ ___
12   11   40   24   18   9   15   5   24   6   15

Answer: By Norse Code

LESSON 1-6 · Practice A
Variables and Algebraic Expressions

Find the value of $n + 3$ for each value of n.

1. $n = 4$ 2. $n = 7$ 3. $n = 0$ 4. $n = 32$

Find the value of $x - 9$ for each value of x.

5. $x = 12$ 6. $x = 57$ 7. $x = 19$ 8. $x = 100$

Find the value of each expression using the given value for each variable.

9. $3n$ for $n = 4$ 10. $x + 8$ for $x = 8$ 11. $9p - 6$ for $p = 2$

12. $n \div 5$ for $n = 35$ 13. $6x + 18$ for $x = 0$ 14. $s - 7$ for $s = 8$

15. $3w + 5$ for $w = 3$ 16. $c - 9$ for $c = 12$ 17. $2a \div 3$ for $a = 6$

18. $y + z$ for $y = 10$ and $z = 20$ 19. $3w - 2v$ for $w = 7$ and $v = 8$

20. $4a \div b$ for $a = 6$ and $b = 4$ 21. $5s + 4t$ for $s = 3$ and $t = 4$

22. The expression $7w$ gives the number of days in w weeks.
Find the value of $7w$ for $w = 20$. How many days are there
in 20 weeks? ___________

23. A cat can run as fast as $m \div 2$ miles per minute in m minutes.
Find the value of $m \div 2$ for $m = 10$. How many miles can a
cat run in 10 minutes? ___________

24. Tyrone works 8 hours a day. You can use the expression $8d$ to
find the total number of hours he works in d days. How many
hours does he work in 5 days? ___________

Holt McDougal Mathematics

Practice B
Variables and Algebraic Expressions

Evaluate $n - 5$ for each value of n.

1. $n = 8$ 2. $n = 121$ 3. $n = 32$ 4. $n = 59$

Evaluate each expression for the given values of the variable.

5. $3n + 15$ for $n = 4$ 6. $h \div 12$ for $h = 60$ 7. $32x - 32$ for $x = 2$

8. $\dfrac{c}{2}$ for $c = 24$ 9. $(n \div 2)5$ for $n = 14$ 10. $8p + 148$ for $p = 15$

11. $e^2 - 7$ for $e = 8$ 12. $3d^2 + d$ for $d = 5$ 13. $40 - 4k^3$ for $k = 2$

14. $2y - z$ for $y = 21$ and $z = 19$ 15. $3h^2 + 8m$ for $h = 3$ and $m = 2$

16. $18 \div a + b \div 9$ for $a = 6$ and $b = 45$ 17. $10x - 4y$ for $x = 14$ and $y = 5$

18. You can find the area of a rectangle with the expression lw where l represents the length and w represents the width. What is the area of the rectangle at right in square feet?

5 ft

2 ft

19. Rita drove an average of 55 mi/h on her trip to the mountains. You can use the expression $55h$ to find out how many miles she drove in h hours. If she drove for 5 hours, how many miles did she drive?

Holt McDougal Mathematics

LESSON 1-6 — Practice C
Variables and Algebraic Expressions

Evaluate each expression for the given values of the variable.

1. $3n + 4n$ for $n = 8$

2. $\dfrac{6s}{5}$ for $s = 25$

3. $q^2 + 5q - 11$ for $q = 4$

4. $\dfrac{350}{d} + 4d + 7$ for $d = 10$

5. $9x + 2x^2 + 2$ for $x = 2$

6. $8m^2 + 7 - 2m$ for $m = 3$

7. $4(h + k)$ for $h = 3$ and $k = 55$

8. $\dfrac{6r}{4} + 5s$ for $r = 8$ and $s = 18$

9. $6a - 2b^2$ for $a = 9$ and $b = 5$

10. $6h - 20g$ for $h = 1{,}500$ and $g = 200$

11. $\dfrac{36}{m^2} + \dfrac{n^2}{4}$ for $m = 6$ and $n = 16$

12. $x^2 - 2x - y^2$ for $x = 15$ and $y = 2$

13. $4d^3 + 6e^2 - \dfrac{8d}{2}$ for $d = 2$ and $e = 3$

14. $\dfrac{5r^2}{4} + \dfrac{4s}{3r}$ for $r = 4$ and $s = 9$

15. You can find the volume of a rectangular prism with the expression $a \cdot b \cdot c$, where a is the length, b is the width, and c is the height of the prism. What is the volume of the prism at the right in cubic inches?

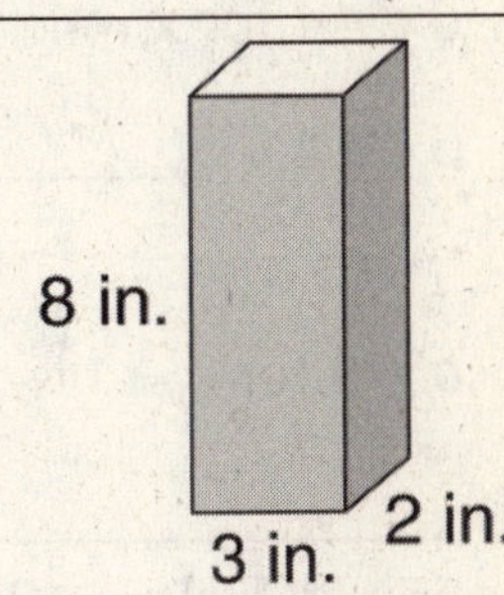

16. You can use the expression $5m$ to find out how many seconds it takes a sound to travel m miles through the air. However, through water, sound takes m seconds to travel m miles. Use the expression $5m - m$ to find out how much longer will it take a sound to travel 8 miles in air than in water.

Holt McDougal Mathematics

<table>
<tr><td>LESSON
1-6</td><td></td></tr>
</table>

Review for Mastery

Variables and Algebraic Expressions

A **variable** is a letter that represents a number than can change in an expression. When you **evaluate** an algebraic expression, you substitute the value given for the variable in the expression.

- Algebraic expression: $x - 3$

  The value of the expression depends on the value of the variable x.

  If $x = 7 \longrightarrow 7 - 3 = 4$

  If $x = 11 \longrightarrow 11 - 3 = 8$

  If $x = 15 \longrightarrow 15 - 3 = 12$

- Evaluate $4n + 1$ for $n = 5$.

  Replace the variable n with 5. $\longrightarrow 4(5) + 1 = 20 + 1 = 21$

Evaluate each expression for the given value.

1. $a + 7$ for $a = 3$

   $a + 7 = 3 + 7 = $ _______

2. $k - 5$ for $k = 13$

   $k - 5 = $ _______ $- 5 = $ _______

3. $y \div 3$ for $y = 6$

   $y \div 3 = $ _______ $\div 3 = $ _______

4. $12 + m$ for $m = 9$

   $12 + m = $ _______ $+ $ _______ $= $ _______

5. $3n - 2$ for $n = 5$

   $3n - 2 = 3(\underline{\quad}) - 2 = $ _______ $- 2 = $ _______

6. $5x + 4$ for $x = 4$

   $5x + 4 = 5(\underline{\quad}) + $ _______ $= $ _______ $+ $ _______ $= $ _______

7. $c - 9$ for $c = 11$

8. $b + 16$ for $b = 4$

9. $a - 4$ for $a = 9$

10. $25 - g$ for $g = 12$

11. $w + 5$ for $w = 2$

12. $3 + s$ for $s = 8$

13. $7q$ for $q = 10$

14. $2y + 9$ for $y = 8$

15. $6x - 3$ for $x = 1$

 Holt McDougal Mathematics

LESSON	**Challenge**
1-6	*What an Expression!*

Complete each table with four expressions that have the same value. Use each given value of *n*.

1.

Expression Value 32	Value of *n* 4
Addition expression:	
Subtraction expression:	
Multiplication expression:	
Division expression:	

2.

Expression Value 96	Value of *n* 12
Addition expression:	
Subtraction expression:	
Multiplication expression:	
Division expression:	

3.

Expression Value 156	Value of *n* 6
Addition expression:	
Subtraction expression:	
Multiplication expression:	
Division expression:	

4.

Expression Value 98	Value of *n* 14
Addition expression:	
Subtraction expression:	
Multiplication expression:	
Division expression:	

5.

Expression Value 57	Value of *n* 12
Addition expression:	
Subtraction expression:	
Multiplication expression:	
Division expression:	

6.

Expression Value 248	Value of *n* 124
Addition expression:	
Subtraction expression:	
Multiplication expression:	
Division expression:	

LESSON 1-6 Problem Solving

Variables and Algebraic Expressions

Write the correct answer.

1. In 2000, people in the United States watched television an average of 29 hours per week. Use the expression $29w$ for $w = 4$ to find out about how many hours per month this is.

2. Find the value of the variable w in the expression $29w$ to find the average number of hours people watched television in a year. Find the value of the expression.

3. The expression $y + 45$ gives the year when a person will be 45 years old, where y is the year of birth. When will a person born in 1992 be 45 years old?

4. The expression $24g$ gives the number of miles Guy's car can travel on g gallons of gas. If the car has 6 gallons of gas left, how much farther can he drive?

Choose the letter for the best answer.

5. Sam is 5 feet tall. The expression $0.5m + 60$ can be used to calculate his height in inches if he grows an average of 0.5 inch each month. How tall will Sam be in 6 months?

   A 56 inches

   B 5 feet 6 inches

   C 63 inches

   D 53 inches

6. The winner of the 1911 Indianapolis 500 auto race drove at a speed of about $s - 56$ mi/h, where s is the 2001 winning speed of about 131 mi/h. What was the approximate winning speed in 1911?

   F 75 mi/h

   G 186 mi/h

   H 85 mi/h

   J 187 mi/h

7. The expression $1{,}587v$ gives the number of pounds of waste produced per person in the United States in v years. How many pounds of waste per person is produced in the United States in 6 years?

   A 1,581 pounds

   B 1,593 pounds

   C 9,348 pounds

   D 9,522 pounds

8. The expression $\$1.25p + \$3.50$ can be used to calculate the total charge for faxing p pages at a business services store. How much would it cost to fax 8 pages?

   F $12.50

   G $4.75

   H $13.50

   J $10.00

  Holt McDougal Mathematics

LESSON 1-6 | Reading Strategies
Focus on Vocabulary

Your age varies from one year to the next. The cost of gasoline can vary from one week to the next. The word **vary** means change. In mathematical expressions, the value of a letter can change, or vary, so letters are called **variables.**

The opposite of vary is to stay **constant.** A constant value never changes.

An algebraic expression is has one or more variables, and may contain constants and operation symbols.

Examples: $4x + 3y - 2^2$ $6p - 3t + 12$

To evaluate an algebraic expression, you need to:

Step 1: Substitute a value for each variable.   **Step 2:** Follow the order of operations.

Evaluate: $4a - 2b + 8$ for $a = 5$ and $b = 3$.
  $4(5) - 2(3) + 8$ ←———————— Substitute 5 for a and 3 for b.
  $20 - 6 + 8$ ←———————— Multiply first.
  $14 + 8$ ←———————— Add and subtract from left to right.
  22

Use this expression for Exercises 1–7: $8 + 4p - 2t$.

1. Write the variables in this expression. _________________________________

2. Rewrite the expression by substituting these values:
   $p = 6$ and $t = 2$.

   ____

3. What operation will you perform first? _________________________________

4. Perform this operation and rewrite the expression.

   ____

5. What operation will you perform next? _________________________________

6. Perform this operation and rewrite the expression. ____________________

7. When $p = 6$ and $t = 2$, what is the solution to the expression $8 + 4p - 2t$?

   ____

      Holt McDougal Mathematics

<table>
<tr><td>**LESSON**
1-6</td><td></td></tr>
</table>

Puzzles, Twisters & Teasers
Movie Math!

Circle words from the list in the word search. Then find an extra word in the word search that best completes the riddle.

expression   substitute   variable   value

constant   evaluate   algebra   algebraic

```
V A L U E S D E X P R E S S I O N
Q A C V B N M V A S D F U B H U I
W L R L K J G A C T U P B V M A L
E G G I L S P L A S H W S I K L V
R E C V A R T U M J U I T C V G N
T B N H T B O A S D F G I P O E U
Y R X O P N L T E R W T T Y T B E
U A V G T M K E U H B J U W Q R D
O I P O I U Y T F I J O T A X A G
P C O N S T A N T D F G E Z X C V
```

This Ron Howard movie was all wet.

___ ___ ___ ___ ___ ___ ___ ___

Holt McDougal Mathematics

LESSON 1-7 Practice A
Translating Words Into Math

Write as an algebraic expression.

1. the sum of *m* and 8

2. the product of 3 and *n*

3. 4 less than *x*

4. the quotient of a number and 12

5. 52 times a number

6. *w* less than 15

7. the sum of 13 and a number

8. the product of 5 and *p*, increased by 10

9. the sum of 15 divided by *b* and 6

10. 12 less than the amount *y* divided by 2

11. 26 increased by 12 times a number _______________________________________

12. the difference of 2 times a number and 6 _______________________________________

13. the product of *h* and 3, increased by 20 _______________________________________

14. 18 less than the product of a number and 4 _______________________________________

15. take away 32 from the product of 6 and a number _______________________________________

16. Used video games cost $25 each. Write an algebraic
    expression to find the cost of *m* video games. _______________________________________

17. Sal earned $740 for *n* weeks of work. Write an algebraic
    expression for the amount he earned each week. _______________________________________

18. At the end of the 2004–2005 NBA season, Reggie Miller
    was the all-time leader in 3-point field goals made.
    He made *n* more field goals than Dale Ellis. Dale Ellis
    made 1,719 3-pointers. Write an algebraic expression to
    find the number of 3-pointers Reggie Miller made. _______________________________________

19. The $2 bill has Thomas Jefferson on the front of it.
    Write an algebraic expression to find out how much money
    v bills with Thomas Jefferson on them would be worth. _______________________________________

  Holt McDougal Mathematics

LESSON 1-7   Practice B
Translating Words Into Math

Write each phrase as an algebraic expression.

1. 125 decreased by a number

2. 359 more than z

3. the product of a number and 35

4. the quotient of 100 and w

5. twice a number, plus 27

6. 12 less than 15 times x

7. the product of e and 4, divided by 12

8. y less than 18 times 6

9. 48 more than the quotient of a number and 64 _____________

10. 500 less than the product of 4 and a number _____________

11. the quotient of p and 4, decreased by 320 _____________

12. 13 multiplied by the amount 60 minus w _____________

13. the quotient of 45 and the sum of c and 17 _____________

14. twice the sum of a number and 600 _____________

15. There are twice as many flute players as there are trumpet players in the band. If there are n flute players, write an algebraic expression to find out how many trumpet players there are. _____________

16. The Nile River is the longest river in the world at 4,160 miles. A group of explorers traveled along the entire Nile in x days. They traveled the same distance each day. Write an algebraic expression to find each day's distance. _____________

17. A slice of pizza has 290 calories, and a stalk of celery has 5 calories. Write an algebraic expression to find out how many calories there are in a slices of pizza and b stalks of celery. _____________

18. Grant pays 10¢ per minute plus \$5 per month for telephone long distance. Write an algebraic expression for m minutes of long-distance calls in one month. _____________

     Holt McDougal Mathematics

LESSON 1-7

Practice C
Translating Words Into Math

Write each phrase as an algebraic expression.

1. the product of 6 and the square of a number _______________________

2. the square of the product of 6 and a number _______________________

3. 4 times the sum of a number and 6,008 _______________________

4. 200 less than half of a number _______________________

5. 3 times the difference of a number squared and 82 _______________________

6. 999 less than 45 increased by the product of a number

   and 85 _______________________

Write a verbal expression for each algebraic expression.

7. $2(4n)$ _______________________

8. $100 - \dfrac{54}{w}$ _______________________

9. $r^2 + 4r + 7$ _______________________

10. $\dfrac{45}{5s^2}$ _______________________

11. An albatross can sleep while flying 25 mi/h. An albatross
    flew 3 miles awake and another n hours asleep at
    25 mi/h. Write an algebraic expression to find the
    distance flown. _______________________

12. You have d dimes, q quarters, and n nickels.
    Write an algebraic expression to find the total
    amount of money. _______________________

13. Four out of every 10 homes in the United States
    have a dog. Write an algebraic expression to find
    out how many dogs there are in h homes. _______________________

14. A waitress who worked k hours earned $32 in
    tips. She gets an additional salary of $4.50 per
    hour. Write an algebraic expression to find the
    amount she earned. _______________________

 Holt McDougal Mathematics

<table>
<tr><td>LESSON
1-7</td><td></td></tr>
</table>

Review for Mastery

Translating Words Into Math

Use the operation clues in a word phrase to translate word phrases into algebraic expressions.

Addition	
add	plus
sum	more than
increased by	

Subtraction	
subtract	minus
difference	less than
decreased by	take away

Multiplication	
times	
multiplied by	
product	

Division	
divided by	
divided into	
quotient	

Write an algebraic expression for the difference of a number and 8.

1. What operation would you choose? _______________________

2. Write an algebraic expression. _______________________

Write an algebraic expression for 3 more than a number.

3. What operation would you choose? _______________________

4. Write an algebraic expression. _______________________

Write an algebraic expression for the quotient of a number and 15.

5. What operation would you choose? _______________________

6. Write an algebraic expression. _______________________

Write an algebraic expression.

7. the product of 12 and a number k _______________________

8. a number d increased by 9 _______________________

9. a number h divided by 4 _______________________

 Holt McDougal Mathematics

Challenge
What's My Equation?

Match each equation with the word problem it represents. Write the equation and corresponding letter. Then write the letter of the equation in the circle that has the problem number. Discover the message formed by the letters.

P $2k = 14$	**D** $\dfrac{p}{3} = 2.50$	**A** $\dfrac{m}{5} = 8$	**U** $3t = 21$
M $a - 18 = 33$	**S** $w + 7 = 25$	**T** $x + 7 = 22$	**H** $n - 12 = 37$

1. Tom has 7 more CDs than Rick does. If Tom has 22 CDs, how many does Rick have? 

2. Maria is 18 years younger than Kim. If Maria is 33, how old is Kim? 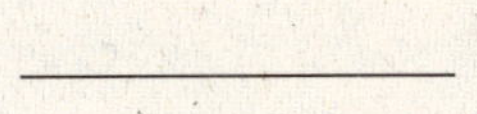 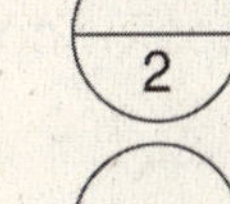

3. Five friends went out for dinner. They shared the cost of the meal equally. If each person paid \$8, what was the total cost of the meal?  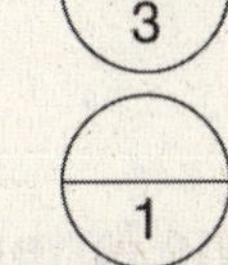

4. Max paid 3 times as much for a tape as his friend did. If Max paid \$21, how much did his friend pay?  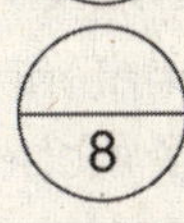

5. Marisa and her two friends share a pizza. The cost of the pizza is shared equally among them. If each person pays \$2.50, how much does the pizza cost?  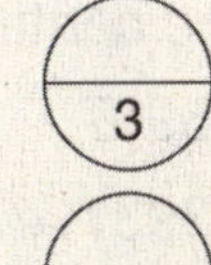

6. Lee has scored twice as many goals as Jiang has. If Lee's goal total is 14, how many goals has Jiang scored? 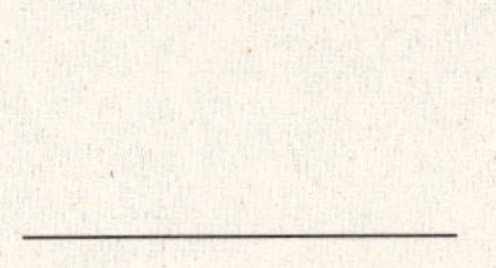 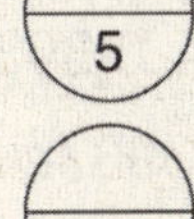

7. Jamal sold 7 more magazine subscriptions than Wayne did. If Jamal sold 25 subscriptions, how many did Wayne sell?  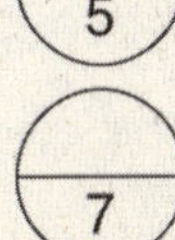

8. Shelly delivered 12 fewer newspapers this week than last week. If she delivered 37 papers this week, how many did she deliver last week? 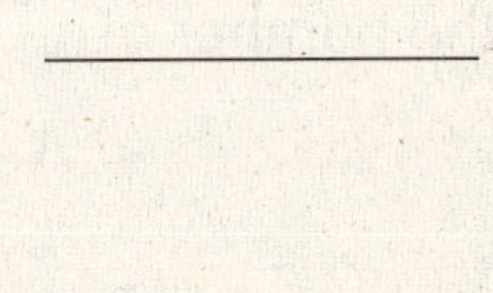 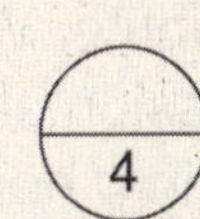 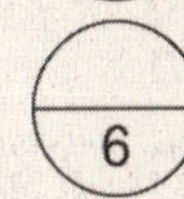

 Holt McDougal Mathematics

LESSON 1-7

Problem Solving

Translating Words into Math

Write the correct answer.

1. Employers in the United States allocate *n* fewer vacation days than the 25 days given by the average Japanese employer. Write an algebraic expression to show the number of vacation days given U.S. workers.

2. There are 112 members in the Somerset Marching Band. They will march in *r* equal rows. Write an algebraic expression for the number of band members in each row.

3. A cup of cottage cheese has 26 grams of protein. Write an algebraic expression for the amount of protein in *s* cups of cottage cheese.

4. Every morning Sasha exercises for 20 minutes. She exercises *k* minutes every evening. Next week she will double her exercise time at night. Write an algebraic expression to show how long Sasha will exercise each day next week.

Choose the letter for the best answer.

5. One centimeter equals 0.3937 inches. Which algebraic expression shows how many inches are in *c* centimeters?

   A $0.3937 + c$

   B $0.3937 \div c$

   C $c \div 0.3937$

   D $0.3937c$

6. In 1957, the Soviet Union launched *Sputnik 1,* the first satellite to orbit Earth. It circled Earth every 1.6 hours for 92 days, then burned up. If the satellite traveled *m* miles per hour, which algebraic expression shows the length of the orbit?

   F $92m$

   G $1.6m$

   H $m \div 1.6$

   J $92 \div m$

7. Gina's heart rate is 70 beats per minute. Which algebraic expression shows the number of beats in *h* hours?

   A $70h$

   B $60h$

   C $4{,}200h$

   D $3{,}600h$

8. The Harris family went on vacation for *w* weeks and 3 days. Which algebraic expression shows the total number of days of their vacation?

   F $7w$

   G $3w$

   H $7w + 3$

   J $3w + 7$

Holt McDougal Mathematics

Name _________________________________ Date _________________ Class _______________

Reading Strategies
Use a Graphic Organizer

Organizing word phrases for the four operations can help you write
algebraic expressions. Study the phrases and the key words
highlighted in this visual map.

<table>
<tr><td>

$r + 12$
- **add** 12 to a number
- a number **plus** 12
- the **sum** of a number and 12
- 12 **more than** a number
- a number increased by 12

</td><td>

$5w$ or $5 \cdot w$
- 5 **times** a number
- 5 **multiplied by** a number
- the **product** of 5 and a number

</td></tr>
<tr><td colspan="2" align="center">**Word Phrases for Algebraic Expressions**</td></tr>
<tr><td>

$n - 8$
- **subtract** 8 from a number
- a number **minus** 8
- 8 **less than** a number
- a number **decreased by** 8
- **take away** 8 from a number

</td><td>

$r + 4$ or $\dfrac{r}{4}$
- 4 **divided into** a number
- the **quotient** of a number and 4
- a number **divided by** 4

</td></tr>
</table>

Write a word phrase for each algebraic expression.

1. $n - 15$ ____

2. $(m + 12) - 3$ ____

3. $2(y + 8)$ ____

4. $5 + \dfrac{w}{3}$ ____

Write an algebraic expression for each word phrase.

5. a number decreased by 9 _______________________________

6. the product of 15 and r _______________________________

7. the quotient of a number and 5 _________________________

8. three times the sum of n and 6 _________________________

     Holt McDougal Mathematics

Puzzles, Twisters & Teasers

Birds of a Feather!

Solve the crossword puzzle. Then use the letters in the shaded boxes to answer the riddle. You'll need to use some letters more than once.

Across

1. something that does not change
2. a number or variable placed to the right of and above another number, variable, or expression
6. putting together
8. the letter in a term
9. a number, a variable, or a product of numbers and variables

Down

1. the number in a term
3. the multiple expressed by an exponent
4. a symbol used for counting
5. group similar objects
7. similar

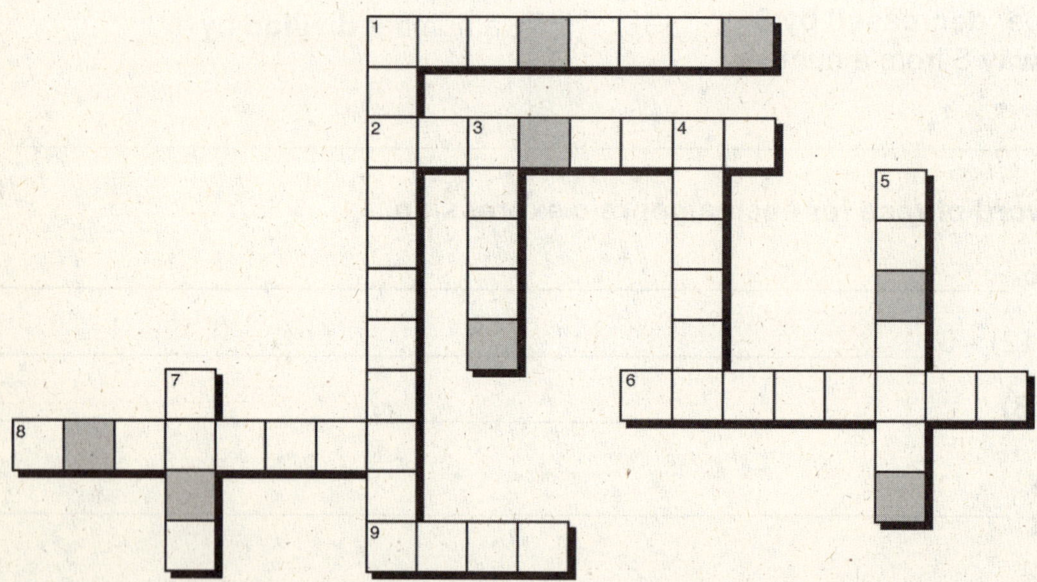

Where do birds invest their money?

In the ______ ______ ______ ______ ______ ______ ______ ______ ______ ______

LESSON 1-8
Practice A
Simplifying Algebraic Expressions

Identify like terms in each list.

1. $6a$ b a 17 $4b$ 32 $17a$

____

2. x x^2 $3x$ 3 $3x^2$ 6

____

3. 2 $6z$ $6z^2$ z $17z$ z^2 3

____

4. m 8 $8m^2$ $8m$ m^2 $12m$ 18

____

5. $2p$ $22p$ $56q$ 12^2 q 34

____

6. d d^2 $15d^2$ $2d$ 4^2 $5d$ 44

____

Combine like terms.

7. $6p^2 + 3p^2$

8. $9x - 6x$

9. $a^2 + b^2 + 2a^2 + 5b^2$

10. $7h^2 + 3 - 2h^2 + 4$

11. $3x + 3y + x + y + z$

12. $5b + 5b + 6b^2 - 10 - 3b$

13. Find the perimeter of the rectangle.
    Combine like terms.

    A $4x + 3y$

    B $8x + 6y$

    C $12xy$

    D $4x^2 + 3y^2$

Holt McDougal Mathematics

Practice B

Simplifying Algebraic Expressions

Identify like terms in each list.

1. $3a$ b^2 b^3 $4b^2$ 4 $5a$

____

2. x x^4 $4x$ $4x^2$ $4x^4$ $3x^2$

____

3. $6m$ $6m^2$ n^2 $2n$ 2 $4m$ $5n$

____

4. $12s$ $7s^4$ $9s$ s^2 5 $5s^4$ 2

____

Simplify. Justify your steps using the Commutative, Associative, and Distributive Properties when necessary.

5. $2p + 22q^2 - p$

____

6. $x^2 + 3x^2 - 4^2$

____

7. $n^4 + n^3 + 3n - n - n^3$

____

8. $4a + 4b + 2 - 2a + 5b - 1$

____

9. $32m^2 + 14n^2 - 12m^2 + 5n - 3$

____

10. $2h^2 + 3g - 2h^2 + 2^2 - 3 + 4g$

____

11. Write an expression for the perimeter of the figure at the right. Then simplify the expression.

____

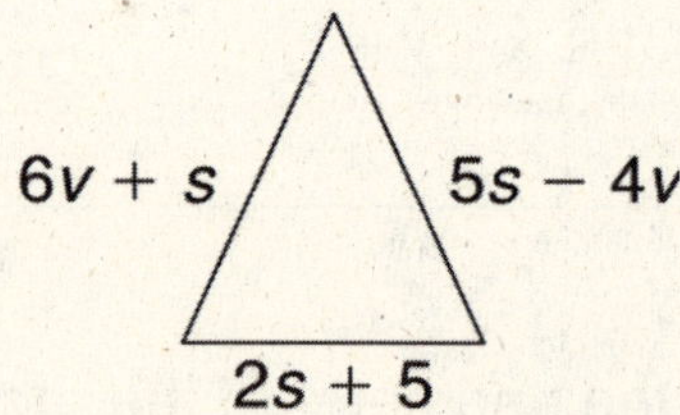

12. Write an expression for the combined perimeters of the figures at the right. Then simplify the expression.

____

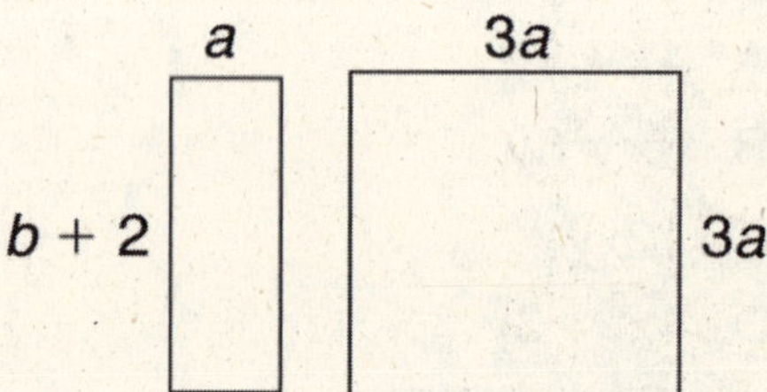

Holt McDougal Mathematics

LESSON 1-8	**Practice C**

Practice C
Simplifying Algebraic Expressions

Simplify each expression. Justify your steps using the Commutative, Associative, and Distributive Properties, when necessary.

1. $8k^2 + 4k - 3k^2 + 3^2 - k + 5$

2. $10x^3 + 5y^2 + 2xy - 4y^2 + 4xy - x^3$

3. $3a + 2b^2 + 6c + a - 2c + b^2 + c$

4. $12x^4 + 6x^2 + 5x^3 - x^2 + 2xy - 8x^4$

5. $9p^6 + q^2 + 6p + 5q^2 + 5p - 5q^2$

6. $h^2 + 4h + 4h^2 - h + 4 + h^2 + 7h$

7. Write an expression that has five terms and simplifies to $5m^3 + 4n$.

8. Write and simplify an expression for the perimeter of the figure to the right.

9. Write an expression to find the combined perimeters of the figures to the right. Then simplify the expression.

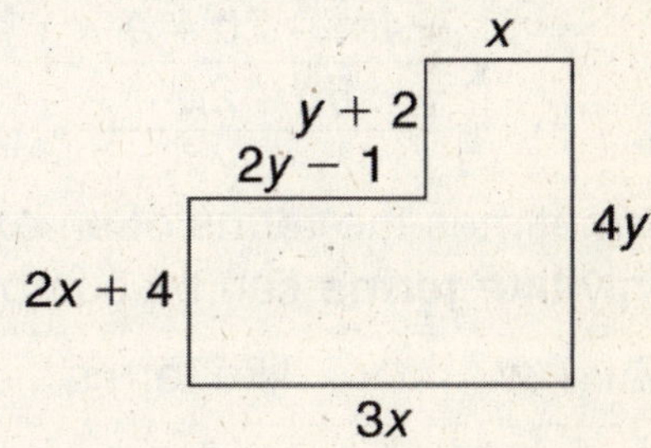

10. Jake scored x points in the first basketball game. He scored 2 fewer points in the next game. His teammate, Jack, scored $2y$ points in the first game and 4 more than twice as many points in the next game. Write and simplify an expression for the total number of points scored by both players.

  Holt McDougal Mathematics

| LESSON 1-8 | # Review for Mastery |

Simplifying Algebraic Expressions

Look at the following expressions: $x = 1x$
$$x + x = 2x$$
$$x + x + x = 3x$$
The numbers 1, 2, and 3 are called **coefficients** of x.

Identify each coefficient.

1. $3n$ ___   2. $7y$ ___   3. m ___   4. 9 ___

An algebraic expression can have terms that are separated by + and –.
In the expression $2x + 5y$, the **terms** are $2x$ and $5y$.

Expression	Terms
$8x + 4y$	$8x$ and $4y$
$m - 3n$	m and $3n$
$4a^2 - 2b + a$	$4a^2$, $2b$, and a
$6d + 2p$	$6d$ and $2p$

Sometimes the terms of an expression can be combined.
Only **like terms** can be combined.

$7w + w$ like terms

$2x - 2y + 2$ unlike terms because x and y are different variables

$8e - 3e + 2e$ like terms

$5d + 25g$ unlike terms because d and g are different variables

To simplify an expression:
Step 1: Combine like terms.
Step 2: Add or subtract the coefficients of the variable.

$$7w + w = 8w$$

$$6y + 1 - 3y = 3y + 1$$

Simplify.

5. $y + 5y$ 6. $9x - 4x$ 7. $5s - 2s$ 8. $3d + 7d$

___________   ___________   ___________   ___________

9. $3b + b + 6$ 10. $8a - a - 3$ 11. $2p + 4p + r$ 12. $9b - 8b + c$

___________   ___________   ___________   ___________

          Holt McDougal Mathematics

LESSON 1-8

Challenge

Matching Terms

Draw a line from each set of terms in Column A to its equivalent combination in Column B. Then circle each letter in Column B that does not have a matching term. Unscramble those letters to answer the riddle.

<table>
<tr><td>

Column A

1. $2x + 7 + 5x - 4 - x$
2. $5 + 7x + 2x - 3 + 6$
3. $x + y + 4x - 3x + 2y + 3y$
4. $3x^2 + 5x - 17 + 6x + 20$
5. $4x + x^2 + 12 - 4 + 2x$
6. $12y + 12x + 12 - 6x + 12$
7. $12y + 4 + x - 7y + 8 + 8x$
8. $5x + x^2 + 2x + 5 - 4 - x^2$
9. $5x^2 + 8x + 7x^2 + 6x$
10. $12x + 6 - 8x - 4x - 3 + 12$
11. $5x + 4 - 3x + 5 + 2x - 9$
12. $4x + 2y + 8 - 3 - y - x$
13. $4x + 5 + 7x + 2y + 2 - y$
14. $2y + 2x + 8 - 6 + x - 2y$
15. $4x + 6y + 6 + 7x + y$
16. $3x^2 + 4x - 2x^2 - 3x + 2x$
17. $8x + 4 - 4 - 4x + x$
18. $y + 5x + 6y + 9 - 6$
19. $x^2 + 3 + 2x^2 + 4 - 7$
20. $5y + 3 + 7x^2 - 2 - x^2 + y$

</td><td>

Column B

A. $5y + 9x + 12$
B. $12y + 6x + 24$
C. 15
D. $9x + 8$
E. 4
F. $6x + 3$
G. $11x + y + 7$
H. $x^2 + 6x + 8$
I. $4x$
J. $3x^2 + 11x + 3$
K. $3x + 2$
L. $3x^2$
M. $6x$
N. $x^2 + 3x$
O. $6x^2 + 6y + 1$
P. $12x^2 + 14x$
Q. $7x + 1$
R. x^2
S. $5x + 7y + 3$
T. 0
U. $2x + 6y$
V. $3x + y + 5$
W. $11x + 7y + 6$
X. $5x$

</td></tr>
</table>

Riddle: What can be a word, a number, a period of time, or a variable?

A ____ ____ ____ ____

   Holt McDougal Mathematics

LESSON 1-8 Problem Solving
Simplifying Algebraic Expressions

Write the correct answer. Use the figures for Problems 1–3.

1. Figure 1 shows the length of each side of a garden. Write and simplify an expression for the perimeter of the garden.

____

2. Figure 2 is a square swimming pool. Write and simplify an expression for the perimeter of the pool.

____

3. Write and simplify an expression for the combined perimeter of the garden and the pool.

____

____

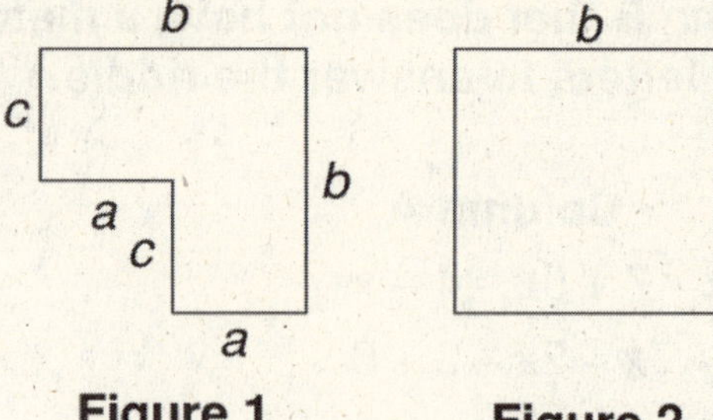

4. The Pantheon in Rome has *n* granite columns in each of 3 rows. Write and simplify an addition expression to show the number of columns. Then evaluate the expression for $n = 8$.

____

____

Choose the letter for the best answer.

5. Which is an expression that shows the earnings of a telemarketer who worked for 23 hours at a salary of *d* dollars per hour?

  A $d + 23$ C $d \div 23$

  B $23d$ D $23 \div d$

6. The minimum wage set in 1997 was $5.15 per hour. Evaluate the expression $40h$ where $h = \$5.15$ to find a worker's weekly salary.

  F $20.60   H $515.00

  G $200   J $206.00

7. What is the perimeter of a triangle with sides the following lengths: $2a + 4c$, $3c + 7$, and $6a - 4$. Simplify the expression.

  A $8a + 11c$

  B $6a + 7c + 3$

  C $8a + 7c + 3$

  D $8a + 7c + 11$

8. A hexagon is a 6-sided figure. Find the perimeter of a hexagon where all of the sides are the same length and the expression $x + y$ represents the length of a side. Simplify the expression.

  F $6x + 6y$

  G $6 + x + y$

  H $6x + y$

  J $6xy$

    Holt McDougal Mathematics

LESSON 1-8
Reading Strategies
Organization Patterns

An algebraic expression is made up of parts called **terms**.

constants	variables	constants and variables
3.2 $\frac{1}{2}$ 12	m s x	$\frac{n}{2}$ $\frac{2}{3}y$ $4x$ $3m^2$

A **coefficient** is a value multiplied by a variable.

Term	Value of Coefficient	Meaning
$7x$	7	$7 \cdot x$
y	1	$1 \cdot y$
$\frac{x}{3}$	$\frac{1}{3}$	$\frac{1}{3} \cdot x$

This expression has 6 terms:

Term	Term	Term	Term	Term	Term
$2x$ +	$5b$ +	7 −	b +	$3x$ +	$2x^2$

- Reorganize the terms: $2x + 3x + 5b - b + 7 + 2x^2$.
- Combine like terms: $5x + 4b + 7 + 2x^2$.

Answer each question.

1. How many terms are there in this expression:
   $6b + b^2 + 5 + 2b - 3f$?

   ____

2. $6b$ and b^2 are unlike terms. Explain why.

   ____

3. How many terms are there in this expression:
   $5a^2 + 6b + a^3 + 2a^2 - 3b - 2$?

   ____

4. Reorganize these terms so like terms are next to each other.

   ____

  Holt McDougal Mathematics

LESSON 1-8	**Puzzles, Twisters & Teasers**
	In Other Words. . .

Write each verbal expression as an algebraic expression. Then use the answer key to solve the riddle.

1. the product of 20 and t _______________ H

2. the sum of 4 times n and 2 _______________ N

3. the product of 7 and p _______________ A

4. the sum of six times n and 1 _______________ B

5. the sum of 5 and a number _______________ I

6. the quotient of n and 8 _______________ C

7. m plus 6 _______________ L

8. t less than 23 _______________ G

9. the quotient of 100 and the amount 6 plus w _______________ S

What's worse than raining cats and dogs?

$\overline{20t}$	$\overline{7p}$	$\overline{5+n}$	$\overline{m+6}$	$\overline{5+n}$	$4n+2$	$23-t$

$\overline{\dfrac{n}{8}}$	$\overline{7p}$	$\overline{6n+1}$	$\overline{\dfrac{100}{(6+w)}}$

 Holt McDougal Mathematics

LESSON 1-9 Practice A
Equations and Their Solutions

Tell if the value of the variable is a solution for the equation.

1. $n = 22$ for $n + 5 = 27$

2. $a = 43$ for $72 - a = 30$

3. $g = 31$ for $19 + g = 40$

4. $z = 87$ for $z - 22 = 65$

5. $f = 15$ for $46 + f = 61$

6. $h = 53$ for $h - 24 = 77$

7. $p = 4$ for $5p = 20$

8. $k = 48$ for $k \div 8 = 8$

9. $m = 12$ for $48 \div m = 4$

10. $x = 8$ for $32x = 264$

11. $v = 5$ for $\dfrac{90}{v} = 15$

12. $s = 7$ for $16s = 112$

13. $m = 2$ for $3m + 4 = 10$

14. $y = 6$ for $3y - 4 = 18$

15. $r = 10$ for $10r - 10 = 90$

16. In the United States in 2005, there were 68 endangered species that were mammals. The number of endangered species that were birds was 9 more than the number of mammal endangered species. Were there 59 or 77 endangered species of birds in the United States?

17. At the Bike and Blade Shop, mountain bikes are on sale for $349. This is $30 more than a racing bike costs. Does the racing bike cost $319 or $379?

18. Which problem situation best matches the equation $2c + 10 = 350$?

Situation A: Austin paid $10 for a new video game console. This is $350 more than 2 times the cost of a used console. How much does a used console cost?

Situation B: Austin paid $350 for a new video game console. This is $10 more than 2 times the cost of a used console. How much does a used console cost?

  Holt McDougal Mathematics

Practice B

Equations and Their Solutions

Determine whether the given value of the variable is a solution.

1. $a = 4$ for $12 - a = 6$

2. $m = 37$ for $23 + m = 60$

3. $x = 6$ for $54 = 9x$

4. $g = 96$ for $\dfrac{g}{4} = 32$

5. $n = 28$ for $n + 44 = 72$

6. $j = 6$ for $84 \div j = 12$

7. $k = 24$ for $3k = 6$

8. $m = 3$ for $42 = m + 39$

9. $y = 8$ for $8y + 6 = 70$

10. $s = 5$ for $18 = 3s - 3$

11. $k = 7$ for $23 - k = 30$

12. $v = 12$ for $84 = 7v$

13. $c = 15$ for $45 - 2c = 15$

14. $x = 10$ for $x + 25 - 2x + 4 = 19$

15. $e = 6$ for $42 = 51 - e$

16. $p = 15$ for $19 = p - 4$

17. Jason and Maya have their own web sites on the Internet. As of last week, Jason's web site had 2,426 visitors. This is twice as many visitors as Maya had. Did Maya have 1,213 visitors or 4,852 visitors to her web site?

18. Which problem situation best matches the equation $3c - 5 = 31$?

Situation A: Rachel had a coupon for $5 off the cost of her order. She ordered 3 large pizzas that each cost the same amount and paid a total of $31. What was the cost of each pizza?

Situation B: Rachel had a coupon for $31 off the cost of her order. She ordered 5 large pizzas that each cost the same amount and paid a total of $3. What was the cost of each pizza?

LESSON 1-9

Practice C

Equations and Their Solutions

Determine whether the given value of the variable is a solution.

1. $a = 15$ for $75 \div a = 5$

2. $x = 90$ for $x \div 9 = 100 - x$

3. $d = 8$ for $875 = 909 - 4d$

4. $x = 32$ for $2x - 25 + x - 70 = 1$

5. $e = 2$ for $e^3 - e^2 = 6e - 8$

6. $b = 4$ for $b^2 + 2b - 3 = 27$

7. $d = 12$ for $4d - 24 - 12 = 0$

8. $r = 9$ for $4r^2 - 19 - 5r = 340$

9. $p = 25$ for $\dfrac{4p}{5} + 2p + 10 = 80$

10. $t = 18$ for $\dfrac{t}{6} + \dfrac{54}{t} = 3$

11. In 1993, there were about 34,000,000 cell phone subscribers in the United States. This is about 125,000,000 fewer subscribers than there were by the year 2003. The equation $34{,}000{,}000 = s - 125{,}000{,}000$ can be used to represent the number of cell phone subscribers in 2003. Were there 91,000,000 or 159,000,000 cell phone subscribers in the United States in 2003?

12. Which problem situation best matches the equation $50n + 15 = 150$?

    Situation A: Hector paid a commission of $150 to buy 15 shares of an Internet stock. Altogether he paid $50. How much did each share of stock cost?

    Situation B: Hector paid a commission of $15 to buy 50 shares of an Internet stock. Altogether he paid $150. How much did each share of stock cost?

Holt McDougal Mathematics

LESSON 1-9

Review for Mastery

Equations and Their Solutions

Number sentences that contain an equal sign (=) are called **equations.**

Equations may be true, or they may be false.

True	False
$3 + 4 = 7$	$3 + 1 = 7$
$8 - 6 = 2$	$8 - 2 = 5$

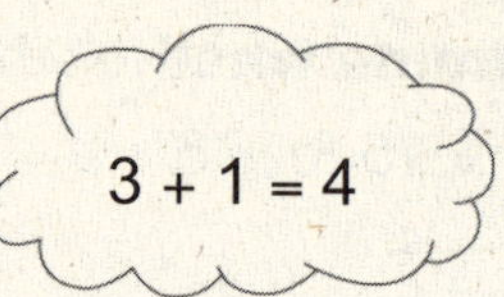

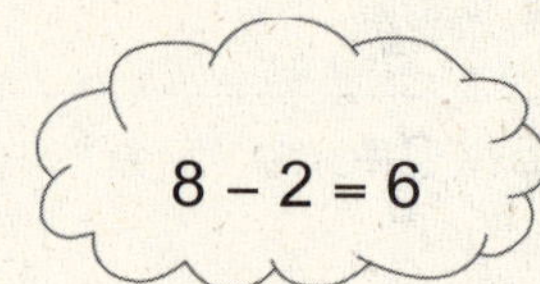

An equation may contain a variable.

variable ➤ $x + 4 = 6$

Whether this equation is true or false depends on the value of x.

You can decide if a number is a *solution* of an equation. Substitute the number for the variable in the equation. If the equation is a true equation, then the number is the **solution.**

Equation: $x + 4 = 6$

Is 2 a solution?

$$x + 4 = 6$$

Substitute 2 for x

$$2 + 4 \overset{?}{=} 6$$ ○ True

$$6 = 6$$ ○

2 is a solution of $x + 4 = 6$.

Is 3 a solution?

$$x + 4 = 6$$

Substitute 3 for x

$$3 + 4 \overset{?}{=} 6$$ ○ False

$$7 \neq 6$$ ○

3 is not a solution of $x + 4 = 6$.

Tell if the number is a solution.

1. Is 3 a solution of $y + 3 = 9$?

2. Is 4 a solution of $n + 6 = 10$?

3. Is 2 a solution of $w - 1 = 1$?

4. Is 1 a solution of $x + 50 = 49$?

5. Is 6 a solution of $c + 23 = 30$?

6. Is 9 a solution of $v - 9 = 0$?

7. Is 20 a solution of $t - 17 = 3$?

8. Is 16 a solution of $12 + a = 24$?

9. Is 25 a solution of $38 - m = 13$?

10. Is 8 a solution of $15 = e + 5$?

Holt McDougal Mathematics

Challenge

LESSON 1-9

The Solution Is BINGO!

Find the solution to each problem or equation. Cross it out on the board below to get BINGO!

1. Is 37, 47, or 67 a solution for $52 = n + 15$?

2. Is 14, 17, or 21 a solution for $8y - 7 = 129$?

3. Is 14, 22, or 24 a solution for $132 - (4x - 5) = 81$?

4. Is 12, 15, or 18 a solution for $3(60 - s) - 2s = 105$?

5. Garret scored 18 points in his last basketball game, which is 6 fewer points than Vince scored. The equation $18 = p - 6$ can be used to represent Vince's points. Did Vince score 12, 24, or 28 points?

6. In 3 years, Sarah's sister will be twice as old as Sarah. If Sarah is now 3 years old, will her sister be 6, 9, or 12 years old in 3 years?

7. The highest recorded temperature in Alaska was 100°F in 1915. This was 56 years before the lowest recorded temperature of –80°F. Was the lowest recorded temperature in 1856, 1956, or 1971?

8. In 2002, Florida had 3,314 public schools. This was 155 more public schools than 3 times the number of public schools in South Carolina during the same year. Did South Carolina have 1,053, 2,849, or 3,159 public schools in 2002?

B	I	N	G	O
17	67	25	37	1971
6	24	47	3,159	22
2,849	1956	FREE	18	14
12	9	16	1,053	1856
28	23	21	19	15

Holt McDougal M

Problem Solving

LESSON 1-9

Equations and Their Solutions

Write the correct answer.

1. The jet airplane was invented in 1939. This is 12 years after the first television was invented. Was television invented in 1927 or 1951?

2. There are three times as many students in the high school as in the junior high school, which has 330 students. Does the high school have 990 students or 110 students?

3. The frigate bird has been recorded at speeds up to 95 mi/h. The only faster bird ever recorded was the spine-tailed swift at 11 mi/h faster. Was the speed of the spine-tailed swift 84 mi/h or 106 mi/h?

4. As of 2004, there were 20.5 million Internet users in Canada. This is 6.6 million more Internet users than there were in Mexico. Were there 27.1 million or 13.9 million Internet users in Mexico?

Choose the letter for the best answer.

5. In the United States, the average school year is 180 days. This is 71 days less than the average school year in China. What is the average school year in China?

   A 251 days

   B 109 days

   C 151 days

   D 271 days

6. The longest suspension bridge in the world is the Akashi Kaikyo Bridge in Japan. Its main span is 1,290 feet longer than a mile. A mile is 5,280 feet. How long is the Akashi Kaikyo bridge?

   F 3,990 feet

   G 6,400 feet

   H 4,049 feet

   J 6,570 feet

7. *Ornithomimus* stood about 6 feet tall and was the fastest dinosaur at a speed of about 50 mi/h. The largest dinosaur, *Seismosaurus,* was 20 times as tall. How tall was *Seismosaurus?*

   A 12 feet

   B 70 feet

   C 120 feet

   D 26 feet

8. Milton collects sports trading cards. He has 80 baseball cards. He has half as many basketball cards as football cards. He has 20 more hockey cards than basketball cards and half as many football cards as baseball cards. How many hockey cards does he have?

   F 20 hockey cards

   G 40 hockey cards

   H 60 hockey cards

   J 80 hockey cards

 Holt McDougal Mathematics

LESSON 1-9 Reading Strategies
Use a Visual Model

The expressions on both sides of the equal sign are equal in
an **equation.**
You can read an equation in math much like you read a sentence.
When you see an equal sign, you read, **"is the same as."**
Balanced scales can help you picture an equation.

Mark has 45 baseball cards.   ⟶   Mark's cards are equal to
This is 8 more than his sister Kathy.   Kathy's cards + 8.

The balanced scales illustrate this equation. ⟶ 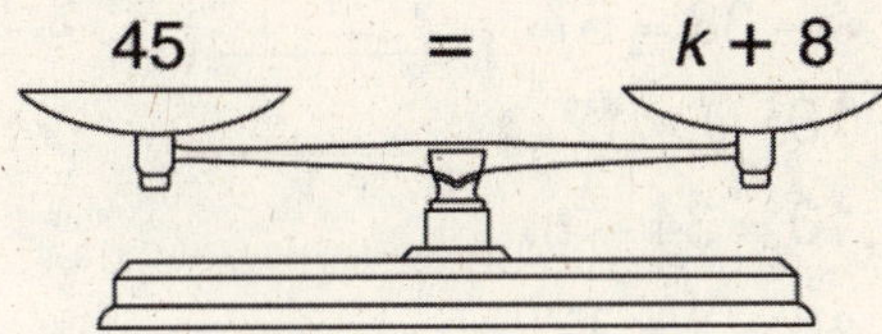

The value of the variable that makes the equation a true sentence
is called the **solution.**

Substitute 40 for k: $45 = k + 8$ Substitute 37 for k: $45 = k + 8$

$\quad\quad\quad\quad\quad 45 \ne 40 + 8$ $45 = 37 + 8$

40 is not a solution.   **37 is the solution.**

Answer each question.

1. What is an equation?

____

____

2. What is the solution of an equation?

____

____

3. Is $3 + 28 \ne 32$ an equation? Why or why not.

____

4. Is $n = 12$ a solution for the equation $43 = n + 31$?
   Why or why not?

____

Determine whether the given value of the variable is a solution.

5. $d + 37 = 54$ for $d = 23$ _______________________________

6. $t - 23 = 42$ for $t = 56$ _______________________________

            Holt McDougal Mathematics

LESSON 1-9
Puzzles, Twisters & Teasers
The Answer Popped Right into My Head!

Find the solution for each equation below. Write the letter on the line above the correct answer at the bottom of the page to solve the riddle.

P $18 = s - 7$ __________

M $x + 3 = 10$ __________

O $12 = t + 9$ __________

Y $s - 38 = 57$ __________

R $16 - j = 12$ __________

I $16 = 34 - m$ __________

U $48 = x + 12$ __________

S $17 + k = 40$ __________

N $24 = 34 - n$ __________

B $p + 18 = 29$ __________

A $47 = v - 6$ __________

G $94 = c + 6$ __________

H $82 = j + 9$ __________

E $a - 15 = 17$ __________

T $15 = k - 9$ __________

What did one firecracker say to the other?

__ __ __ __ __ __ __
7   95   25   3   25   18   23

__ __ __ __ __ __ __ __ __ __
11   18   88   88   32   4   24   73   53   10

__ __ __ __ __ __ __
95   3   36   4   25   3   25

LESSON 1-10

Practice A

Solving Equations by Adding or Subtracting

Match each equation in Column A with its correct solution in Column B.

Column A	Column B	Column A	Column B
1. $n - 16 = 8$	A. $n = 12$	10. $x - 12 = 13$	L. $x = 14$
2. $5 = n - 7$	B. $n = 13$	11. $x + 8 = 40$	M. $x = 17$
3. $12 + n = 25$	C. $n = 17$	12. $34 = 16 + x$	N. $x = 18$
4. $n - 17 = 11$	D. $n = 24$	13. $x + 5 = 19$	P. $x = 25$
5. $n + 18 = 35$	E. $n = 27$	14. $4 + x = 52$	Q. $x = 32$
6. $7 = n - 28$	F. $n = 28$	15. $12 + x = 50$	R. $x = 33$
7. $n - 12 = 40$	G. $n = 35$	16. $15 = x - 2$	S. $x = 38$
8. $24 = n - 25$	H. $n = 49$	17. $52 = x + 9$	T. $x = 43$
9. $46 = n + 19$	J. $n = 52$	18. $x - 11 = 22$	U. $x = 48$

19. Chris has 55 baseball trading cards. He has 17 more cards than his sister Sara has. Write and solve an equation to find how many trading cards Sara has.

____

20. In 2000, Sammy Sosa hit 50 home runs. His home run total was 23 runs fewer than the number of home runs that Barry Bonds hit the next year. Write and solve an equation to find how many home runs Barry Bonds hit in 2001.

____

 Holt McDougal Mathematics

| LESSON 1-10 | **Practice B** |

Solving Equations by Adding or Subtracting

Solve each equation. Check your answer.

1. $33 = y - 44$

2. $r - 32 = 77$

3. $125 = x - 29$

4. $k + 18 = 25$

5. $589 + x = 700$

6. $96 = 56 + t$

7. $a - 9 = 57$

8. $b - 49 = 254$

9. $987 = f - 11$

10. $32 + d = 1{,}400$

11. $w - 24 = 90$

12. $95 = g - 340$

13. $e - 35 = 59$

14. $84 = v + 30$

15. $h + 15 = 81$

16. $110 = a + 25$

17. $45 + c = 91$

18. $p - 29 = 78$

19. $56 - r = 8$

20. $39 = z + 8$

21. $93 + g = 117$

22. The Morales family is driving from Philadelphia to Boston. So far, they have driven 167 miles. This is 129 miles less than the total distance they must travel. How many miles is Philadelphia from Boston?

23. Ron has $1,230 in his savings account. This is $400 more than he needs to buy a new big screen TV. Write and solve an equation to find out how much the TV costs.

Holt McDougal Mathematics

Practice C
Solving Equations by Adding or Subtracting

Solve each equation. Check your answer.

1. $b - 32 = 15$

2. $e - 43 = 121$

3. $601 = x - 24$

4. $m + 45 = 123$

5. $314 + z = 350$

6. $840 = 45 + f$

7. $d - 67 = 23$

8. $w + 233 = 319$

9. $91 = x + 52$

10. $150 + y = 879$

11. $k - 32 = 217$

12. $408 = s - 129$

13. $108 = j - 24$

14. $1{,}204 = w + 389$

15. $p - 167 = 321$

16. In 2003, *USA Today* was the leading U.S. daily newspaper, with a circulation of 2,154,539. This was 63,477 more than the second leading daily newspaper, the *Wall Street Journal*. Write and solve an equation to find the circulation of the *Wall Street Journal* in 2003.

17. Mrs. Baker's class has been raising money for a local charity. At the end of last week, they had collected $238. By the end of this week, they had a total of $419. Write and solve an equation to find the amount collected this week.

18. Andy Green broke the sound barrier on land on October 15, 1997, in Black Rock Desert, Nevada. The speed of sound was recorded at about 751 mi/h. This was about 12 mi/h less than the land speed record that he set that day. Write and solve an equation to find Andy Green's land speed record.

　　　　　　　　　　　　　　　Holt McDougal Mathematics

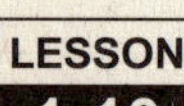

LESSON 1-10

Review for Mastery

Solving Equations by Adding or Subtracting

Solving an equation is like balancing a scale. If you add the same weight to both sides of a balanced scale, the scale will remain balanced. You can use this same idea to solve an equation.

Think of the equation $x - 7 = 12$ as a balanced scale. The equal sign keeps the balance.

$$
\boxed{-7 + 7 = 0}
\begin{aligned}
x - 7 &= 12 \\
x - 7 + \mathbf{7} &= 12 + \mathbf{7} \quad &\text{Add 7 to both sides.} \\
x + 0 &= 19 \quad &\text{Combine like terms.} \\
x &= 19
\end{aligned}
$$

When you solve an equation, the idea is to get the variable by itself. What you do to one side of the equation, you must do to the other side.

- To solve a subtraction equation, use addition.
- To solve an addition equation, use subtraction.

Solve and check: $y + 8 = 14$.

$$
\boxed{+8 - 8 = 0}
\begin{aligned}
y + 8 &= 14 \\
y + 8 - \mathbf{8} &= 14 - \mathbf{8} \quad &\text{Subtract 8 from both sides.} \\
y + 0 &= 6 \quad &\text{Combine like terms.} \\
y &= 6
\end{aligned}
$$

Check:
$$
\begin{aligned}
y + 8 &= 14 \quad &\text{To check, substitute 6 for } y. \\
6 + 8 &\overset{?}{=} 14 \\
14 &= 14
\end{aligned}
$$

A true sentence, $14 = 14$, means the solution is correct.

Solve and check.

1. $x - 2 = 8$

  $x - 2 + \underline{\quad} = 8 + \underline{\quad}$

  $x - 0 = \underline{\quad}$

2. $b + 5 = 11$

  $b + 5 - \underline{\quad} = 11 - \underline{\quad}$

  $b + 0 = \underline{\quad}$

3. $n + 8 = 11$

4. $y - 6 = 2$

5. $a - 9 = 4$

6. $m + 2 = 18$

_______________ _______________ _______________ _______________

Holt McDougal Mathematics

LESSON 1-10

Challenge
Equation Maker

Use each term once to make up one addition and one subtraction equation, then solve the equations.

1. *m, n*, 12, 6, 54, 9

____

____

2. *x, y*, 7, 15, 32, 45

____

____

3. *p, q*, 19, 44, 72, 8

____

____

4. *a, b*, 67, 102, 6, 8

____

____

5. *c, d*, 11, 12, 18, 35

____

____

6. *s, t*, 115, 123, 32, 0

____

____

7. *w, y*, 1, 2, 3, 4

____

____

8. *n, p*, 6, 22, 99, 400

____

____

9. *e, f*, 52, 4, 75, 18

____

____

10. *g, h*, 61, 88, 94, 117

____

____

11. *k, l*, 302, 54, 115, 79

____

____

12. *r, s*, 90, 14, 71, 15

____

____

13. *u, v*, 8, 12, 37, 44

____

____

14. *x, y*, 198, 0, 231, 4

____

____

   Holt McDougal Mathematics

LESSON 1-10

Problem Solving
Solving Equations by Adding or Subtracting

Write the correct answer.

1. In an online poll, 1,927 people voted for Coach as the best job at the Super Bowl. The job of Announcer received 8,055 more votes. Write and solve an equation to find how many votes the job of Announcer received.

2. In 2005, the largest bank in the world was UBS, Switzerland, with $1,533 billion in assets. This was $49 billion more than the largest bank in the United States, Citigroup. Write and solve an equation to find Citigroup's assets.

3. The two smallest countries in the world are Vatican City and Monaco. Vatican City is 1.37 square kilometers smaller than Monaco, which is 1.81 square kilometers in area. What is the area of Vatican City?

4. The Library of Congress is the largest library in the world. It has 29 million books, which is 10 million more than the National Library of Canada has. How many books does the National Library of Canada have?

Choose the letter for the best answer.

5. The first track on Sean's new CD has been playing for 55 seconds. This is 42 seconds less than the time of the entire first track. How long is the first track on this CD?

   A 37 seconds   C 97 seconds

   B 63 seconds   D 93 seconds

6. There are 45 students on the school football team. This is 13 more than the number of students on the basketball team. How many students are on the basketball team?

   F 58 students   H 32 students

   G 48 students   J 42 students

7. A used mountain bike costs $79.95. This is $120 less than the cost of a new one. If c is the cost of the new bike, which equation can you use to find the cost of a new bike?

   A $79.95 = c + 120$

   B $120 = 79.95 - c$

   C $79.95 = c - 120$

   D $120 = 79.95 + c$

8. The goal of the School Bake Sale is to raise $125 more than last year's sale. Last year the Bake Sale raised $320. If it reaches its goal, how much will the Bake Sale raise this year?

   F $445

   G $195

   H $525

   J $425

Holt McDougal Mathematics

LESSON 1-10 Reading Strategies
Follow a Procedure

In order to solve an equation, you must find the **solution.** The solution is the value that makes the equation true. To solve an equation, you need to get the variable by itself on one side of the equal sign.

- If you have an addition equation, you must subtract to get the variable by itself.

- If you have a subtraction equation, you must add to get the variable by itself.

Example:

$z + 12 = 32$ ⟵ To get z by itself, subtract 12.

$z + 12 - \mathbf{12} = 32 - \mathbf{12}$ ⟵ Rewrite the equation to show that 12 is subtracted from both sides.

$z = 20$ ⟵ This is the solution after subtracting 12 from both sides.

Check by using 12 in place of z.

$20 + 12 \stackrel{?}{=} 32$

$32 = 32$, so $z = 20$ is the correct solution.

Example:

$27 = x - 8$ ⟵ To get x by itself, add 8.

$27 + \mathbf{8} = x - 8 + \mathbf{8}$ ⟵ Rewrite the equation to show that 8 is added to both sides.

$35 = x$ ⟵ This is the solution after adding 8 to both sides.

Check by using 35 in place of x.

$27 \stackrel{?}{=} 35 - 8$

$27 = 27$, so $x = 35$ is the correct solution.

Use $m + 17 = 43$ for Exercises 1–4.

1. What operation is shown in this equation? _______________________________

2. What operation will you use to get m by itself? _______________________________

3. Rewrite the equation showing subtracting from both sides of the equation. _______________________________

4. What is the value of m? _______________________________

  Holt McDougal Mathematics

LESSON 1-10

Puzzles, Twisters & Teasers
Clean Solutions!

Find the solution for each equation below. Write the letter of each variable on the line above the correct answer at the bottom of the page to solve the riddle.

1. $y - 35 = 17$ ___________

2. $h - 40 = 26$ ___________

3. $a + 16 = 43$ ___________

4. $110 = e + 66$ ___________

5. $97 = w - 44$ ___________

6. $n - 8 = 3$ ___________

7. $356 = g - 218$ ___________

8. $652 + t = 800$ ___________

9. $16 = o - 124$ ___________

10. $63 + m = 903$ ___________

11. $k + 18 = 98$ ___________

12. $d - 27 = 54$ ___________

13. $c - 50 = 23$ ___________

14. $l - 62 = 937$ ___________

Why did the robber take a shower?

| 66 | 44 | 141 | 27 | 11 | 148 | 44 | 81 | 148 | 140 |

| 840 | 27 | 80 | 44 | 27 | 73 | 999 | 44 | 27 | 11 |

| 574 | 44 | 148 | 27 | 141 | 27 | 52 |

LESSON 1-11

Practice A
Solving Equations by Multiplying or Dividing

Solve.

1. $16 = n \div 2$

2. $\dfrac{e}{10} = 8$

3. $25 = \dfrac{x}{6}$

4. $18 = \dfrac{d}{3}$

5. $a \div 12 = 7$

6. $30 = b \div 4$

Solve and check.

7. $7w = 49$

8. $75 = 3x$

9. $60 = 12p$

10. $77 = 11m$

11. $4h = 48$

12. $9y = 54$

13. $2x = 30$

14. $45 = 5s$

15. $6z = 42$

16. The Fruit Stand charges $0.50 each for navel oranges. Kareem paid $4.00 for a large bag of navel oranges. How many did he buy?

17. Jenny can type at a speed of 80 words per minute. It took her 20 minutes to type a report. How many words was the report?

18. At the local gas station, regular unleaded gasoline is priced at $1.10 per gallon. If it cost $16.50 to fill a car's gas tank, how many gallons of gasoline did the tank hold?

Holt McDougal Mathematics

<table>
<tr><td>LESSON
1-11</td><td></td></tr>
</table>

Practice B
Solving Equations by Multiplying or Dividing

Solve each equation. Check your answer.

1. $68 = \dfrac{r}{4}$

2. $k \div 24 = 85$

3. $255 = \dfrac{x}{4}$

____

4. $42 = w \div 18$

5. $\dfrac{a}{15} = 22$

6. $82 = b \div 5$

____

7. $\dfrac{c}{7} = 9$

8. $28 = z \div 3$

9. $\dfrac{y}{12} = 10$

____

Solve each equation. Check your answer.

10. $52w = 364$

11. $41x = 492$

12. $410 = 82p$

____

13. $35d = 735$

14. $195 = 65h$

15. $4k = 140$

____

16. $110 = 5e$

17. $27a = 216$

18. $96 = 12n$

____

19. Ashley earns $5.50 per hour babysitting. She wants to buy a CD player that costs $71.50, including tax. How many hours will she need to work to earn the money for the CD player?

____

20. A cat can jump the height of up to 5 times the length of its tail. How high can a cat jump if its tail is 13 inches long?

____

 Holt McDougal Mathematics

LESSON 1-11 Practice C

Solving Equations by Multiplying or Dividing

Solve each equation. Check your answer.

1. $765 = \dfrac{n}{12}$

2. $\dfrac{m}{9} = 26$

3. $\dfrac{a}{12} = 14$

4. $3g = 165$

5. $308 = 44b$

6. $27e = 405$

Translate each sentence into an equation. Then solve the equation.

7. The product of a number w and 145 is 725. _______________________________

8. The quotient of a number f and 21 is 14. _______________________________

9. A number b times 23 equals 253. _______________________________

10. A number k divided by 15 equals 47. _______________________________

11. A number n multiplied by 12 is 84. _______________________________

12. Ten divided into a number m equals 54. _______________________________

13. About three tons of ore must be mined and processed to produce a single ounce of gold. How many tons of ore are required to produce a pound of gold?

____

14. It costs about $2,931 per hour to operate a Boeing 757 airplane. Find the cost to operate a Boeing 757 during a 5-hour flight.

____

15. Each person in the United States eats an average of 23 quarts of ice cream per year. At this rate, the seventh-grade class will eat about 1,794 quarts of ice cream this year. How many students are in the seventh-grade class?

____

16. Jumbo shrimp sell for $14.99 per pound in a local supermarket. One customer spent $74.95 on shrimp for a dinner party. How much shrimp did this customer purchase for the party?

____

 Holt McDougal Mathematics

Review for Mastery
LESSON 1-11

Solving Equations by Multiplying or Dividing

When you solve an equation, you must get the variable by itself. Remember, what you do to one side of an equation, you must do to the other side.

- To solve a division equation, multiply both sides of the equation by the same number.

Solve and check: $\dfrac{a}{3} = 4$.

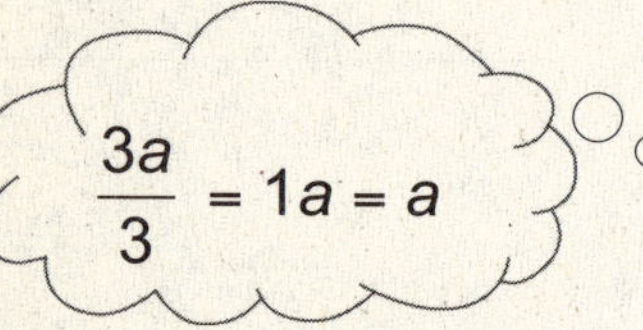

$$\dfrac{a}{3} = 4$$

$$(3)\dfrac{a}{3} = 4(3)$$

$$a = 12$$

Check: $\dfrac{a}{3} = 4$

$$\dfrac{12}{3} \overset{?}{=} 4$$

$$4 \overset{?}{=} 4 \quad ✔$$

Solve and check.

1. $\dfrac{x}{6} = 3$

2. $\dfrac{s}{8} = 8$

3. $\dfrac{c}{10} = 7$

4. $\dfrac{n}{3} = 12$

_______   _______   _______   _______

- To solve a multiplication equation, divide both sides of the equation by the same number.

Solve and check: $5k = 30$.

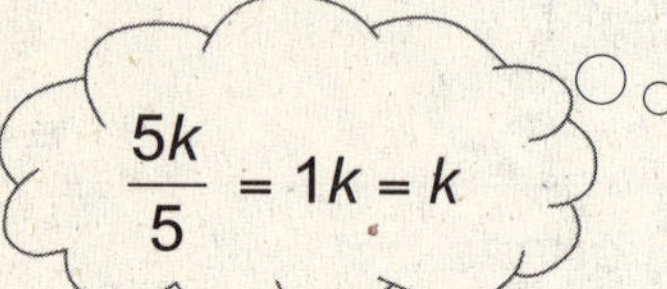

$$5k = 30$$

$$\dfrac{5k}{5} = \dfrac{30}{5}$$

$$k = 6$$

Check: $5k = 30$

$$5(6) \overset{?}{=} 30$$

$$30 \overset{?}{=} 30 \quad ✔$$

Solve and check.

5. $2w = 16$

6. $4b = 24$

7. $9z = 45$

8. $10m = 40$

_______   _______   _______   _______

Holt McDougal Mathematics

<table>
<tr><td>LESSON
1-11</td><td>

Challenge
Shape Up

</td></tr>
</table>

Find the value of each shape in Exercises 1–8. Then use the values to answer the questions below.

1. 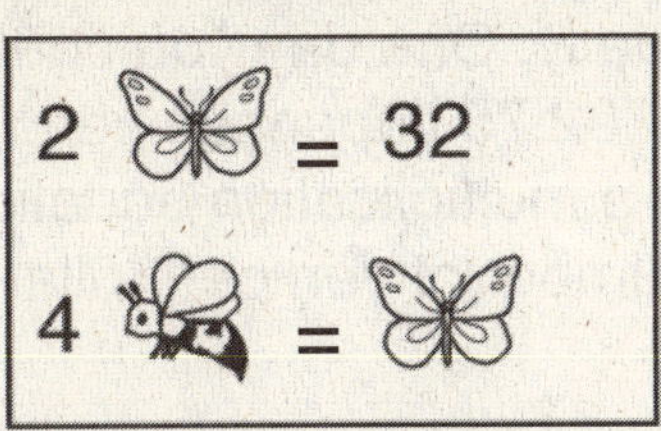

2. 

3. 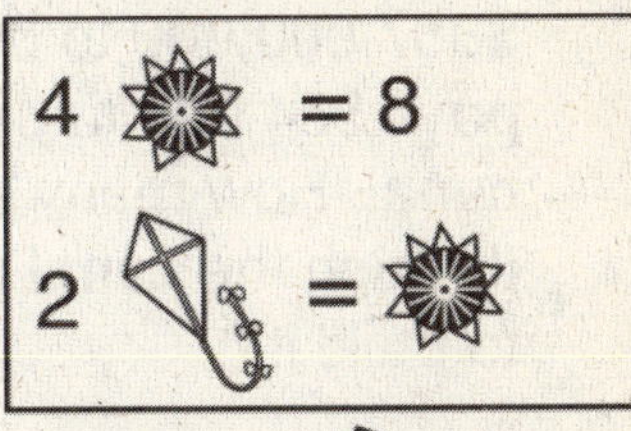

4. 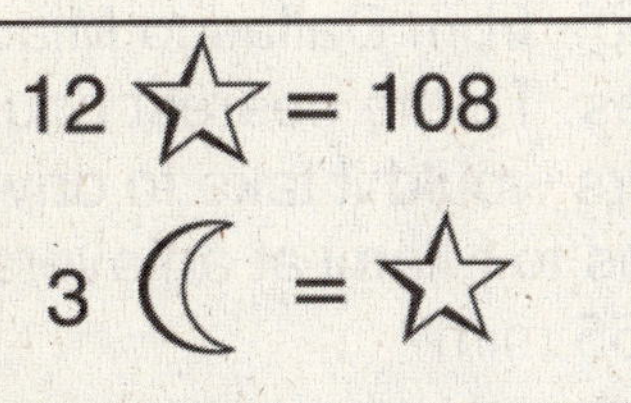

5. 

6.

7.

8. 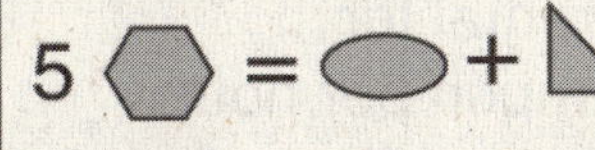

9. The British colonies in North America eventually became the United States. What was their population in 1610?

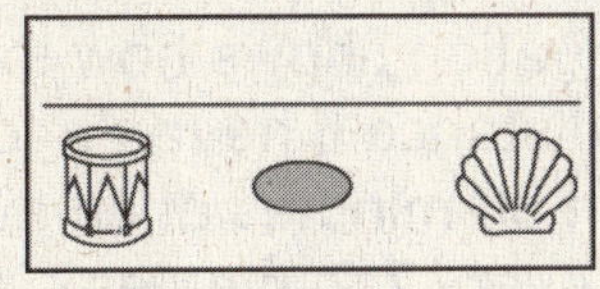

10. What was the population of the United States in 2000?

Holt McDougal Mathematics

Problem Solving

Solving Equations by Multiplying or Dividing

Write the correct answer.

1. The Panama Canal cost $387,000,000 to build. Each ship pays $34,000 to pass through the canal. How many ships had to pass through the canal to pay for the cost to build it?

2. The rate of exchange for currency changes daily. One day you could get $25 for 3,302.75 Japanese yen. Write and solve a multiplication equation to find the number of yen per dollar on that day.

3. Franklin D. Roosevelt was in office as president for 12 years. This is three times as long as Jimmy Carter was president. Write and solve an equation to show how long Jimmy Carter was president.

4. The mileage from Dallas to Miami is 1,332 miles. To the nearest hour, how many hours would it take to drive from Dallas to Miami at an average speed of 55 mi/h?

Choose the letter for the best answer.

5. The total bill for a bike rental for 8 hours was $38. How much per hour was the rental cost?

   A $8 per hour

   B $4.75 per hour

   C $30 per hour

   D $5.25 per hour

6. If a salesclerk earns $5.75 per hour, how many hours per week does she work to earn her weekly salary of $207?

   F 30 hours

   G 32 hours

   H 36 hours

   J 4 hours

7. At a cost of $0.07 per minute, which equation could you use to find out how many minutes you can talk for $3.15?

   A $0.07 \div m = $3.15

   B $3.15 \cdot m = $0.07

   C $0.07m = $3.15

   D $0.07 \div $3.15 = m

8. Which equation shows how to find a runner's distance if he ran a total of m miles in 36 minutes at an average of a mile every 7.2 minutes?

   F $36 \div m = 7.2$

   G $7.2 \div m = 36$

   H $36m = 7.2$

   J $7.2 \div 36 = m$

Holt McDougal Mathematics

LESSON
1-11

Reading Strategies
Follow a Procedure

The opposite of multiplication is division: $\longrightarrow$ $12 \cdot 3 = 36$, and $36 \div 3 = 12$

The opposite of division is multiplication: $\longrightarrow$ $48 \div 12 = 4$, and $4 \cdot 12 = 48$

From these examples you can see that:

division "undoes" multiplication, and **multiplication "undoes" division.**

To solve multiplication and division equations:

- Get the variable by itself on one side of the equation.

- Keep the equation in balance by using the same operation on both sides.

Example:

$84 = 7x$ $\longleftarrow$ Get the variable by itself. This is a multiplication equation, so divide to "undo" the multiplication.

$\dfrac{84}{7} = \dfrac{7x}{7}$ $\longleftarrow$ Rewrite the equation to show that both sides are divided by 7.

$12 = x$ $\longleftarrow$ This is the solution after dividing both sides by 7.

Check using 12 in place of x:

$84 \overset{?}{=} 7(12)$

$84 = 84$, so $x = 12$ is the solution.

Example:

$\dfrac{m}{15} = 8$ $\longleftarrow$ Get the variable by itself. Multiply to "undo" division.

$\dfrac{m}{15} \cdot 15 = 8 \cdot 15$ $\longleftarrow$ Rewrite the equation to show that both sides are multiplied by 15.

$m = 120$ $\longleftarrow$ This is the solution after multiplying both sides by 15.

Check by using 120 in place of m.

$\dfrac{120}{15} \overset{?}{=} 8$

$8 = 8$, so $m = 120$ is the solution.

Use $108 = 9y$ for Exercises 1–3.

1. What operation will you use to solve the equation?

____

2. Rewrite the equation using the inverse operation on both sides.

____

3. What is the value of y?

____

   Holt McDougal Mathematics

Puzzles, Twisters & Teasers

LESSON 1-11

Doctor! Doctor!

Find the solution for each equation below. Write the letter of each variable on the line above the correct answer at the bottom of the page to solve the riddle.

1. $a \div 25 = 4$ _______________

2. $w \times 18 = 18$ _______________

3. $r \div 8 = 5$ _______________

4. $3h = 96$ _______________

5. $72 = 8d$ _______________

6. $12 = y \div 4$ _______________

7. $17 = n \div 8$ _______________

8. $85 = 17o$ _______________

9. $3e = 63$ _______________

10. $9 = u \div 3$ _______________

11. $6b = 222$ _______________

12. $7m = 84$ _______________

13. $150 = 3t$ _______________

14. $9s = 99$ _______________

Patient: Doctor! Doctor! I feel like an umbrella!

Doctor: ___ ___ ___   ___ ___ ___
     1   32   48   48   5   27

  ___ ___ ___ ___   ___ ___
  12 27 11 50   37 21

  ___ ___ ___ ___ ___   ___ ___ ___
  27 136 9 21 40   50 32 21

  ___ ___ ___ ___ ___ ___ ___
  1 21 100 50 32 21 40

Holt McDougal Mathematics

Answers

Practice A

1. 19, 22, 25

   add 3 to each number to get the next number

2. 37, 26, 15

   subtract 11 from each number to get the next number

3. 32, 64, 128

   multiply each number by 2 to get the next number

4. 21, 28, 36

   add 1 more than you did the time before

5. 

   a circle and two triangles

6. 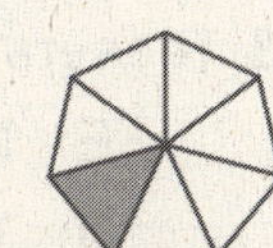

   shade every other triangle in a counterclockwise direction.

7.

Figure	1	2	3
Number of Dots	2	6	10

   18

Practice B

1. 25, 21, 17

   Subtract 4 from each number to get the next number.

2. 70, 80, 92

   Add 2 more than you did the time before.

3. 20, 10, 5

   Divide each number by 2 to get the next number.

4. 88, 104, 120

   Add 16 to each number to get the next number.

5. 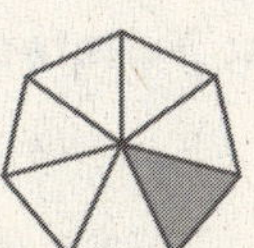

   Alternate triangles and squares with circles between them.

6.

   Rotate the figure 90° clockwise each time.

7.

Figure	1	2	3
Number of Dots	1	4	9

   25

Practice C

1. 164, 187, 210

   Add 23 to each number to get the next number.

2. 58; 37, 30

   Subtract 7 from each number to get the next number.

3. 36; 2,916

   Multiply each number by 3 to get the next number.

4. 50, 49; 35

   Subtract 1 more than you did the time before.

5.

6. 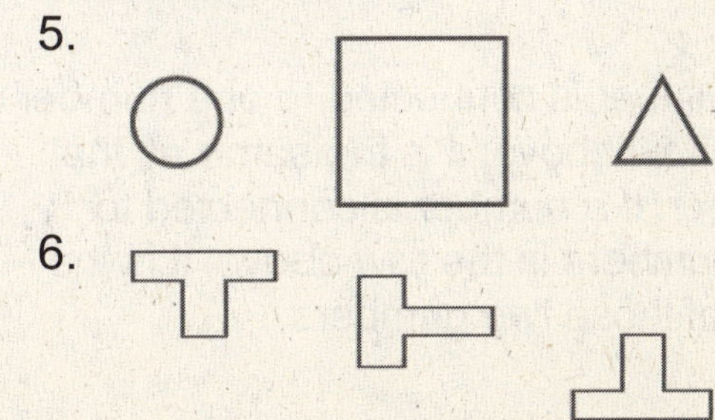

 Holt McDougal Mathematics

7.

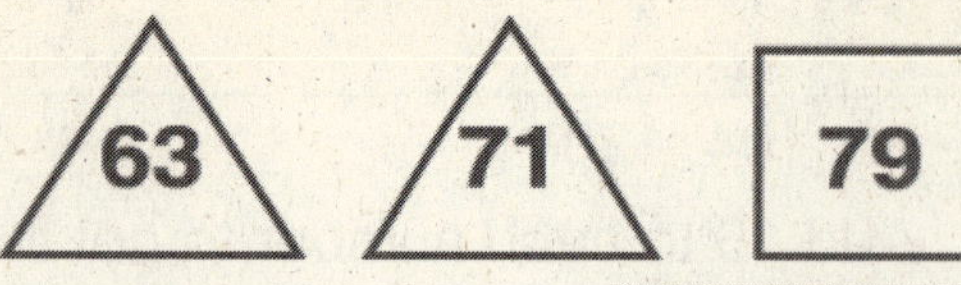

8.

Figure	1	2	3
Number of Squares	2	6	12

30

Review for Mastery

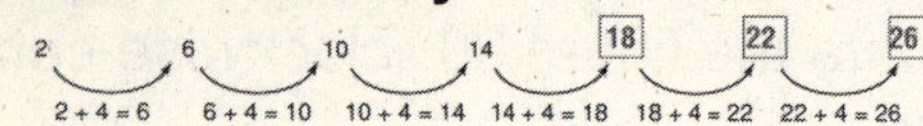

1. 19, 16, 13;
   3;
   3; 19;
   19; 3; 16;
   16; 3; 13

2. 78, 96, 114;
   18;
   18; 78;
   78; 18; 96;
   96; 18; 114

3. 36, 44, 52; Add 8

4. 48, 96, 192; Multiply by 2

5.   

6. 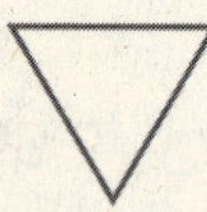 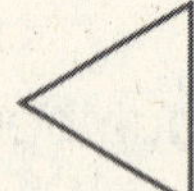 

Challenge

1. 2; 4; 8; 16; 32; the sum doubles for each row.

2. 512

3. If a number is connected to one number in the row above; it is the same as that number. If a number is connected to two numbers in the row above, it is the sum of those two numbers.

4. 1, 6, 15, 20, 15, 6, 1
   1, 7, 21, 35, 35, 21, 7, 1
   1, 8, 28, 56, 70, 56, 28, 8, 1
   1, 9, 36, 84, 126, 126, 84, 36, 9, 1

5. 512; it matches.

Problem Solving

1. 8:36 A.M.

2. No; the pattern is multiples of 8, and 174 is not a multiple of 8.

3. 70 gallons

4. sphere, pyramid, sphere, sphere, cube

5. D   6. H

7. A   8. G

Reading Strategies

1. Each number in the pattern is 9 less than the number it comes after.

2. 28, 19, 10

3. The figure rotates 90° clockwise each time.

Puzzles; Twisters & Teasers

T: 11   E: 112

H: 70   J: 3

W: 64   K: 22

R: 512   N: 8

S: 35   O: 102

L: 486

WHEN THE JOKER IS WILD

LESSON 1-2

Practice A

1. 16   2. 2; 2; 2; 8

3. 6; 6; 36   4. 81

5. 64   6. 243

7. 1   8. 100

9. 81   10. 9

11. 32   12. 5^2

Holt McDougal Mathematics

13. 3^1

14. 2^3

15. 4^0

16. 9^2

17. 4^3

18. 8^2

19. 3^2

20. 6^2

21. 2^4

22. 3^3

23. 20^2

24. 2^3; $8

25. 25 miles

Practice B

1. 25

2. 16

3. 27

4. 49

5. 256

6. 144

7. 1,000

8. 11

9. 1

10. 400

11. 216

12. 343

13. 4^2

14. 25^1

15. 10^2

16. 5^3

17. 2^5

18. 3^5

19. 30^2

20. 11^2

21. 60^2

22. 4^4

23. 8^3

24. 14^2

25. 4^3 stamps; 64 stamps

26. 81 minutes

Practice C

1. 6,561

2. 16,807

3. 2,500

4. 6,561

5. 27,000

6. 248,832

7. >

8. <

9. >

10. <

11. <

12. =

13. 7^3

14. 5^4

15. 12^3

16. 15^2

17. 100^3

18. 35^2

19. 9^4

20. 3^7

21. 20^3

22. 7^4 postcards; 2,401 postcards

Review for Mastery

1. 4; 4; 4

2. 1; 1; 1; 1; 1; 1

3. 25

4. 8

5. 27

6. 36

7. 64

8. 4

9. 125

10. 16

11. 6^3

12. 4^2

13. 2^3

14. 3^2

15. 9^2

16. 3^3

17. 7^2

Challenge

1. $1,600

2. $2,200

3. $1,200

4. $2,000

5. $2,025

6. $10,125

7. $300

8. 6 times

9. $19,200

10. $468

11. 3 times

12. $12,636

Problem Solving

1. 4,096 cells

2. 13,156,488

3. Prize B: $215,233.59

4. 81 great-great-grandchildren

5. C

6. G

7. D

8. G

Reading Strategies

1. five to the third power

2. five is a factor 3 times: $5 \times 5 \times 5$

3. 125

4. Sample answer: 5^3 means $5 \times 5 \times 5 = 125$, and 3^5 means $3 \times 3 \times 3 \times 3 \times 3 = 243$.

5. $32 = 2 \times 2 \times 2 \times 2 \times 2$

6. $32 = 2^5$

7. $6 \times 6 \times 6 = 216$

8. $216 = 6^3$

Puzzles, Twisters & Teasers

E: 81

T: 7^2

Holt McDougal Mathematics

K: 1,000

B: 9^2

I: 64

D: 25

W: 18

C: 5^2

A: 3^3

H: 1

V: 144

R: 4^4

O: 6^3

U: 125

S: 36

BECAUSE IT HAD A HARD DRIVE

LESSON 1-3

Practice A

1. C
2. G
3. A
4. H
5. 2,300
6. 150,000
7. 3,000
8. 28,000
9. 13,200
10. 201,000
11. 45,600
12. 1,080,000
13. 5.6×10^4
14. 3.06×10^5
15. 8.0×10^6
16. 7.2×10^6
17. 1.4×10^7
18. 4.1×10^1
19. 2.144×10^6
20. 2.03×10^6
21. 2.3×10^4
22. 36,000,000

Practice B

1. 6,000
2. 220
3. 800
4. 18
5. 7,000
6. 25,000
7. 30,000
8. 180,000
9. 840,000
10. 31,500
11. 210,000
12. 1,004,000
13. 17,640
14. 856
15. 4,055,000
16. 7,160,000
17. 3.4×10^4
18. 7.7×10^3
19. 2.1×10^6
20. 4.04×10^5
21. 2.1×10^7
22. 6.12×10^2
23. 3.001×10^6
24. 6.213×10^5
25. 3.17×10^4
26. 142,000,000
27. 1,306,000,000
26. 4.8×10^6

Practice C

1. 5,000
2. 47,100
3. 395
4. 20,000,000
5. 7,025
6. 5,700,000
7. 662,500
8. 9,010,000,000
9. 2.5×10^4
10. 9.9×10^3
11. 9.7×10^6
12. 9.56×10^9
13. 2.3×10^{10}
14. 1.1×10^2
15. 3.019×10^7
16. 7.355×10^5
17. 4
18. 830,000
19. 1.12
20. 4.1
21. 77,000,000
22. 6
23. 1.4325×10^5
24. 403,000,000
25. Brazil
26. 3.7×10^7

Review for Mastery

1. 3; 84,000
2. 5; 6,100,000
3. 22,000,000
4. 753,000
5. 82,500
6. 1,230
7. 3;
   5; 1; ^{3}
8. 6;
   1; 84; ^{6}
9. 6.41×10^5
10. 4.73×10^4
11. 8.25×10^6
12. 7.03×10^5

Challenge

1. 6.4×10^4 bytes
2. 4×10^7 bytes
3. 1×10^9 bytes
4. 2.5×10^{11} bytes
5. 1.44×10^6 bytes
6. 1×10^5 disks
7. 6.5×10^8 bytes
8. 6×10^9 bytes

Problem Solving

1. 1.5×10^8 km
2. 4,500,000,000 km
3. $7,600,000,000,000; 7.6 \times 10^{12}$

Holt McDougal Mathematics

4. Canada   5. C

6. J   7. A

8. H

Reading Strategies

1. 5 times

2. $2.8 \times 10 \times 10 \times 10 \times 10 \times 10$

3. 5 places; 4   4. 280,000

Puzzles, Twisters & Teasers

1. 4   2. 5

3. 5.92   4. 1.68

5. 8   6. 2.44; 7

SPACE CHIPS

LESSON 1-4

Practice A

1. B   2. J

3. A   4. G

5. C   6. H

7. 7   8. 38

9. 12   10. 1

11. 48   12. 27

13. 80   14. 1

15. 14   16. $8

17. $52

Practice B

1. 69   2. 219

3. 21   4. 53

5. 21   6. 10

7. 33   8. 88

9. 28   10. 0

11. 57   12. 39

13. 39   14. 11

15. 46   16. 7

17. 18   18. 12

19. 18   20. 95

21. 19   22. 44

23. 0   24. 17

25. $10   26. $22

Practice C

1. 195   2. 154

3. 51   4. 71

5. 524   6. 19

7. 81   8. 130

9. 85   10. 1

11. 38   12. 8

13. 7   14. 18

15. 12   16. $18 + 2(1 + 3^2)$

17. $(4 \cdot 2 - 2)^2 \div 9 + 2$

18. $3^3 - (9 \cdot 2 + 1)$ 19. $2^3 - (3 \cdot 5 - 8)$

20. $35 + 4^2 - (6 - 3)$

21. $6 \cdot 7 - 3 \cdot (4 + 1)$

22. $93   23. $19

Review for Mastery

1. 48; 46   2. 5; 25

3. 45; 9   4. 13

5. 3   6. 8

7. 0   8. 9

9. 1   10. 17

11. 4   12. 1

Challenge

1. $313   2. $342

3. $268   4. $180

5. $7(5 \cdot 12) + 11(4 \cdot 12)$; $948

6. $7(5 \cdot 12) + 11(4 \cdot 12) + 35$; $983

7. $12(5 \cdot 12) + 18(4 \cdot 12) - [7(5 \cdot 12) + 11(4 \cdot 12) + 35]$; $601

8. $577

Problem Solving

1. $35 \cdot 2.10 - 12$; $61.50

2. $3 \cdot 3.50 + 4 \cdot 3$; $22.50

3. $324 per week

4. She has $31 left.

5. B   6. F

7. C   8. J

      Holt McDougal Mathematics

Reading Strategies

1. divide → multiply → subtract → add
2. 7
3. parentheses → multiply → add → exponents → subtract
4. 15
5. parentheses → divide → add → exponent → subtract
6. 45

Puzzles, Twisters & Teasers

1. T	2. F
3. F	4. T
5. F	6. T
7. T	8. F
9. T	

YOUR FEET OFF THE FLOOR

LESSON 1-5

Practice A

1. Identity Property
2. Associative Property
3. Commutative Property
4. Identity Property

Answer for the below are given from left to right

5. 20; 48; Associative Property; Add
6. 5; 5; 100; 17; 1,700; Commutative Property; Associative Property; Multiply
7. 7; 10; 40; 28; 68
8. Possible work shown.
   3 • (20 + 8);
   (3 • 20) + (3 • 8);
   60 + 24; 84

Practice B

1. Commutative Property
2. Identity Property
3. Associative Property
4. Associative Property

5. Possible answers are given.
   9 • 4 • 50; Commutative Property
   9 • (4 • 50); Associative Property
   9 • 200; Multiply.
   1,800; Multiply.
6. Possible answers are given.
   (45 + 33) + 7; Commutative Property
   45 + (33 + 7); Associative Property
   45 + 40; Add.
   85; Add.
7. Possible work shown.
   3 • (20 + 6);
   (3 • 20) + (3 • 6);
   60 + 18;
   78
8. Possible work shown.
   (20 – 2)9;
   (20 • 9) – (2 • 9);
   180 – 18;
   162

Practice C

1. 0; Identity Property
2. 7; Distributive Property
3. 45; Commutative Property
4. 14; Associative Property
5. Possible answers are given.
   (7 • 40) • 5; Commutative Property
   7 • (40 • 5); Associative Property
   7 • 200; Multiply.
   1,400; Multiply.
6. Possible answers are given.
   98 + 15 + 85; Commutative Property
   98 + (15 + 85); Associative Property
   98 + 100; Add.
   198
7. 7 • (40 + 3);
   (7 • 40) + (7 • 3);
   280 + 21;
   301

     Holt McDougal Mathematics

8. (600 – 3) • 4;
   (600 • 4) – (3 • 4);
   2,400 – 12;
   2,388

Review for Mastery

1. 45; Commutative
   45; 25; Associative
   70;
   109

2. 4; 7; Commutative
   25; 4; 7; Associative
   100; 7;
   700;

3. 8;
   10; 8;
   50; 40;
   90; Distributive

4. 1;
   30; 1;
   180; 6;
   174; Distributive

Challenge

612	621	374
306	612	540
86	612	612

768	864	864
768	256	864
256	300	864

Problem Solving

1. Possible answer: $45 + $17 + $25; use
   the Commutative and Associative
   properties to rewrite the expression as
   ($45 + $25) + $17, then add mentally.

2. Possible answer: (12 • 8) • 25,
   12 • (8 • 25); 12 • (8 • 25) is easier to
   multiply, because 12 • 200 is easier to
   multiply than 96 • 25; 2,400

3. 270 mi

4. Possible answer: Yes. Since 50 + 14 =
   64, 6(64) = 6(50 + 14) = 6(50) + 6(14).

5. A   6. J

7. C   8. H

Reading Strategies

1. 16 and 14
2. 16 + (14 + 39); Commutative Property
3. (16 + 14) + 39; Associative Property
4. 69   5. 35 and 5
6. 47 + 35 + 5; Commutative Property
7. 47 + (35 + 5); Associative Property
8. 87

Puzzles, Twisters, & Teasers

1. Associative Property
2. Identity Property
3. Distributive Property
4. Commutative Property
5. (5 + 7) + 6   6. 5(70) + 5(6)
7. 70 + 11   8. 7 • (5 • 6)
9. 5(7) + 5(6)   10. 5(70) – 5(6)
11. (5 • 70) • 6

THEY KEEP THEIR EYES PEELED

LESSON 1-6

Practice A

1. 7   2. 10
3. 3   4. 35
5. 3   6. 48
7. 10   8. 91
9. 12   10. 16
11. 12   12. 7
13. 18   14. 1
15. 14   16. 3
17. 4   18. 30
19. 5   20. 6
21. 31   22. 140 days
23. 5 miles   24. 40 hours

Practice B

1. 3   2. 116
3. 27   4. 54

Holt McDougal Mathematics

5. 27

6. 5

7. 32

8. 12

9. 35

10. 268

11. 57

12. 80

13. 8

14. 23

15. 43

16. 8

17. 120

18. 10 square feet

19. 275 miles

Practice C

1. 56

2. 30

3. 25

4. 82

5. 28

6. 73

7. 232

8. 102

9. 4

10. 5,000

11. 65

12. 191

13. 78

14. 23

15. 48 cubic inches

16. 32 seconds longer

Review for Mastery

1. 10

2. 13; 8

3. 6; 2

4. 12; 9; 21

5. 5; 15; 13

6. 4; 4; 20; 4; 24

7. 2

8. 20

9. 5

10. 13

11. 7

12. 11

13. 70

14. 25

15. 3

Challenge

1. $n + 28$; $36 - n$; $8n$; $128 \div n$

2. $n + 84$; $108 - n$; $8n$; $1{,}152 \div n$

3. $n + 150$; $162 - n$; $26n$; $936 \div n$

4. $n + 84$; $112 - n$; $7n$; $1{,}372 \div n$

5. $n + 45$; $69 - n$; $4.75n$; $684 \div n$

6. $n + 124$; $372 - n$; $2n$; $30{,}752 \div n$

Problem Solving

1. about 116 hours

2. 52; 1,508 hours

3. 2037

4. 144 miles

5. C

6. F

7. D

8. H

Reading Strategies

1. p and t

2. $8 + 4(6) - 2(2)$

3. multiplication

4. $8 + 24 - 4$

5. addition

6. $32 - 4$

7. 28

Puzzles, Twisters & Teasers

V	A	L	U	E	S	D	E	X	P	R	E	S	S	I	O	N
Q	A	C	V	B	N	M	V	A	S	D	F	U	B	H	U	I
W	L	R	L	K	J	G	A	C	T	U	P	B	V	M	A	L
E	G	G	I	L	S	P	L	A	S	H	W	S	I	K	L	V
R	E	C	V	A	R	T	U	M	J	U	I	T	C	V	G	N
T	B	N	H	T	B	O	A	S	D	F	G	I	P	O	E	U
Y	R	X	O	P	N	L	T	E	R	W	T	T	Y	T	B	E
U	A	V	G	T	M	K	E	U	H	B	J	U	W	Q	R	D
O	I	P	O	I	U	Y	T	F	I	J	O	T	A	X	A	G
P	C	O	N	S	T	A	N	T	D	F	G	E	Z	X	C	V

SPLASH

LESSON 1-7

Practice A

1. $m + 8$

2. $3n$

3. $x - 4$

4. $a \div 12$

5. $52k$

6. $15 - w$

7. $13 + e$

8. $5p + 10$

9. $15 \div b + 6$

10. $y \div 2 - 12$

11. $26 + 12s$

12. $2a - 6$

13. $3h + 20$

14. $4n - 18$

15. $6z - 32$

16. $25m$

17. $740 \div n$

18. $1{,}719 + n$

19. $2v$

Holt McDougal Mathematics

Practice B

1. $125 - n$
2. $z + 359$
3. $35f$
4. $100 \div w$
5. $2r + 27$
6. $15x - 12$
7. $4e \div 12$
8. $18 \cdot 6 - y$
9. $m \div 64 + 48$
10. $4t - 500$
11. $p \div 4 - 320$
12. $13(60 - w)$
13. $45 \div (c + 17)$
14. $2(d + 600)$
15. $n \div 2$
16. $4{,}160 \div x$
17. $290a + 5b$
18. $0.10m + 5$

Practice C

1. $6n^2$
2. $(6n)^2$
3. $4(r + 6{,}008)$
4. $y \div 2 - 200$ or $\dfrac{y}{2} - 200$
5. $3(n^2 - 82)$
6. $(45 + 85e) - 999$

Possible answers given.

7. twice the product of a number and 4
8. 100 less the quotient of 54 and w
9. r squared plus 4 times r, plus 7
10. 45 divided by the product of 5 and a number squared
11. $25n + 3$
12. $0.10d + 0.25q + 0.05n$
13. $\dfrac{4h}{10}$
14. $4.50k + 32$

Review for Mastery

1. subtraction
2. $n - 8$
3. addition
4. $n + 3$
5. division
6. $n \div 15$
7. $12k$
8. $d + 9$
9. $h \div 4$

Challenge

1. $x + 7 = 22$; T
2. $a - 18 = 33$; M
3. $\dfrac{m}{5} = 8$; A
4. $3t = 21$; U
5. $\dfrac{p}{3} = 2.50$; D
6. $2k = 14$; P
7. $w + 7 = 25$; S
8. $n - 12 = 37$; H

MATH ADDS UP

Problem Solving

1. $25 - n$
2. $112 \div r$
3. $26s$
4. $20 + 2k$
5. D
6. G
7. C
8. H

Reading Strategies

1. possible answer: 15 less than n
2. possible answer: 3 less than the sum of m and 12
3. possible answer: twice the sum of y and 8
4. possible answer: 5 increased by the quotient of a number and 3
5. $n - 9$
6. $15r$
7. $p \div 5$
8. $3(n + 6)$

Puzzles, Twisters & Teasers

Across
1. constant
2. exponent
6. grouping
8. variable
9. term

Down
1. coefficient
3. power
4. number
5. combine
7. like

STORK MARKET

LESSON 1-8

Practice A

1. $6a$, a, and $17a$; b and $4b$; 17 and 32

Holt McDougal Mathematics

2. x^2 and $3x^2$; x and $3x$; 3 and 6

3. $6z^2$ and z^2; z, $6z$, and $17z$; 2 and 3

4. $8m^2$ and m^2; m, $8m$, and $12m$; 8 and 18

5. $2p$ and $22p$; $56q$ and q; 12^2 and 34

6. d, $2d$, and $5d$; d^2 and $15d^2$; 4^2 and 44

7. $9p^2$ 8. $3x$

9. $3a^2 + 6b^2$ 10. $5h^2 + 7$

11. $4x + 4y + z$ 12. $6b^2 + 7b - 10$

13. B

Practice B

1. $3a$ and $5a$; b^2 and $4b^2$

2. x and $4x$; x^4 and $4x^4$; $4x^2$ and $3x^2$

3. $6m$ and $4m$; $2n$ and $5n$

4. $12s$ and $9s$; $7s^4$ and $5s^4$; 5 and 2

5. $p + 22q^2$ 6. $4x^2 - 16$

7. $n^4 + 2n$ 8. $2a + 9b + 1$

9. $20m^2 + 14n^2 + 5n - 3$

10. $7g + 1$ 11. $2v + 8s + 5$

12. $14a + 2b + 4$

Practice C

1. $5k^2 + 3k + 14$ 2. $9x^3 + y^2 + 6xy$

3. $4a + 3b^2 + 5c$

4. $4x^4 + 5x^3 + 5x^2 + 2xy$

5. $9p^6 + q^2 + 11p$ 6. $6h^2 + 10h + 4$

7. Possible answer: $3m^3 + 5n + 4m^3 - 2m^3 - n$

8. $6x + 7y + 5$ 9. $12a + 14b + 12$

10. $2x + 6y + 2$

Review for Mastery

1. 3   2. 7

3. 1   4. 0

5. $6y$ 6. $5x$

7. $3s$ 8. $10d$

9. $4b + 6$ 10. $7a - 3$

11. $6p + r$ 12. $b + c$

Challenge

1. F   2. D

3. U   4. J

5. H   6. B

7. A   8. Q

9. P   10. C

11. I   12. V

13. G   14. K

15. W   16. N

17. X   18. S

19. L   20. O

TERM

Problem Solving

1. $2a + 2b + 2c$ 2. $4b$

3. $2a + 6b + 2c$

4. $n + n + n = 3n$; $3(8) = 24$; 24 columns

5. B   6. J

7. C   8. F

Reading Strategies

1. 5 terms

2. In b^2, b is raised to the second power, but it is not in $6b$.

3. 6

4. $a^3 + 5a^2 + 2a^2 + 6b - 3b - 2$

Puzzles, Twisters & Teasers

1. $20t$ 2. $4n + 2$

3. $7p$ 4. $6n + 1$

5. $5 + n$ 6. $\dfrac{n}{8}$

7. $m + 6$ 8. $23 - t$

9. $\dfrac{100}{(6 + w)}$

HAILING CABS

Holt McDougal Mathematics

LESSON 1-9

Practice A

1. yes	2. no
3. no	4. yes
5. yes	6. no
7. yes	8. no
9. yes	10. no
11. no	12. yes
13. yes	14. no
15. yes	

16. 77 endangered species

17. $319

18. Situation B

Practice B

1. no	2. yes
3. yes	4. no
5. yes	6. no
7. no	8. yes
9. yes	10. no
11. no	12. yes
13. yes	14. yes
15. no	16. no

17. 1,213 visitors

18. Situation A

Practice C

1. yes	2. yes
3. no	4. yes
5. yes	6. no
7. no	8. no
9. yes	10. no

11. 159,000,000 cell phone subscribers

12. Situation B

Review for Mastery

1. no	2. yes
3. yes	4. no
5. no	6. yes
7. yes	8. no
9. yes	10. no

Challenge

1. 37	2. 17
3. 14	4. 15
5. 24	6. 12
7. 1971	8. 1,053

B	I	N	G	O
~~17~~	67	25	~~37~~	~~1971~~
6	~~24~~	47	3,159	~~22~~
2,849	1956	FREE	18	~~14~~
~~12~~	9	16	~~1,053~~	1856
28	23	21	19	~~15~~

Problem Solving

1. 1927	2. 990 students

3. 106 mi/h

4. 13.9 million Internet users

5. A	6. J
7. C	8. G

Reading Strategies

1. a mathematical sentence with expressions of the same value on both sides of the equal sign

2. sample answer: the value of the variable that makes the expressions equal

3. no; The expressions are not equal in value.

4. yes; $43 = 12 + 31$

5. $d = 23$ is not a solution.

6. $t = 56$ is not a solution.

Puzzles, Twisters & Teasers

P: 25

M: 7

O: 3

Y: 95

R: 4

Holt McDougal Mathematics

I: 18

U: 36

S: 23

N: 10

B: 11

A: 53

G: 88

H: 73

E: 32

T: 24

MY POP IS BIGGER THAN YOUR POP

LESSON 1-10

Practice A

1. D	2. A
3. B	4. F
5. C	6. G
7. J	8. H
9. E	10. P
11. Q	12. N
13. L	14. U
15. S	16. M
17. T	18. R

19. $55 = 17 + n$; $n = 38$

20. $50 = h - 23$; $h = 73$

Practice B

1. $y = 77$	2. $r = 109$
3. $x = 154$	4. $k = 7$
5. $x = 111$	6. $t = 40$
7. $a = 66$	8. $b = 303$
9. $f = 998$	10. $d = 1,368$
11. $w = 114$	12. $g = 435$
13. $e = 94$	14. $v = 54$
15. $h = 66$	16. $a = 85$
17. $c = 46$	18. $p = 107$
19. $r = 48$	20. $z = 31$
21. $g = 24$	

22. The total distance is 296 miles.

23. $1,230 = t + 400$; The TV costs $830.

Practice C

1. $b = 47$	2. $e = 164$
3. $x = 625$	4. $m = 78$
5. $z = 36$	6. $f = 795$
7. $d = 90$	8. $w = 86$
9. $x = 39$	10. $y = 729$
11. $x = 249$	12. $s = 537$
13. $j = 132$	14. $w = 815$
15. $p = 488$	

16. $2,154,539 = c + 63,477$; $c = 2,091,062$

17. $\$238 + m = \$419$; $m = \$181$

18. $751 = r - 12$; $r = 763$ mi/h

Review for Mastery

1. 2; 2; 10	2. 5; 5; 6 6
3. $n = 3$	4. $y = 8$
5. $a = 13$	6. $m = 16$

Challenge

Possible answers given.

1. $n - 12 = 54$; $n = 66$; $m + 6 = 9$; $m = 3$

2. $x - 15 = 32$; $x = 47$; $y + 7 = 45$; $y = 38$

3. $p + 19 = 72$; $p = 53$; $q - 8 = 44$; $q = 52$

4. $a - 102 = 6$; $a = 108$; $b + 8 = 67$; $b = 59$

5. $c - 18 = 35$; $c = 53$; $d + 11 = 12$; $d = 1$

6. $s + 115 = 123$; $s = 8$; $t - 32 = 0$; $t = 32$

7. $w - 2 = 1$; $w = 3$; $y + 3 = 4$; $y = 1$

8. $n - 6 = 400$; $n = 406$; $p + 22 = 99$; $p = 77$

9. $e + 4 = 75$; $e = 71$; $f - 18 = 52$; $f = 70$

10. $g - 117 = 61$; $g = 178$; $h + 88 = 94$; $h = 6$

11. $k - 54 = 115$; $k = 169$; $l + 79 = 302$; $l = 223$

12. $r + 15 = 90$; $r = 75$; $s - 14 = 71$; $s = 85$

Holt McDougal Mathematics

13. $u - 37 = 44$; $u = 81$; $v + 8 = 12$; $v = 4$

14. $x - 198 = 4$; $x = 202$; $y + 0 = 231$;
    $y = 231$

Problem Solving

1. $1{,}927 = v - 8{,}055$; $v = 9{,}982$;
   9,982 votes

2. $1{,}533 = a + 49$; $a = 1{,}484$;
   $1,484 billion

3. 0.44 square km

4. 19 million books

5. C   6. H

7. C   8. F

Reading Strategies

1. addition   2. subtraction

3. $m + 17 - 17 = 43 - 17$

4. $m = 26$

Puzzles, Twisters & Teasers

1. $y = 52$ 2. $h = 66$

3. $a = 27$ 4. $e = 44$

5. $w = 141$ 6. $n = 11$

7. $g = 574$ 8. $t = 148$

9. $o = 140$ 10. $m = 840$

11. $k = 80$ 12. $d = 81$

13. $c = 73$ 14. $l = 999$

HE WANTED TO MAKE A CLEAN GET
   AWAY

LESSON 1-11

Practice A

1. $n = 32$ 2. $e = 80$

3. $x = 150$ 4. $d = 54$

5. $a = 84$ 6. $b = 120$

7. $w = 7$ 8. $x = 25$

9. $p = 5$ 10. $m = 7$

11. $h = 12$ 12. $y = 6$

13. $x = 15$ 14. $s = 9$

15. $z = 7$ 16. 8 oranges

17. 1,600 words

18. 15 gallons of gasoline

Practice B

1. $r = 272$ 2. $k = 2{,}040$

3. $x = 1{,}020$ 4. $w = 756$

5. $a = 330$ 6. $b = 410$

7. $c = 63$ 8. $z = 84$

9. $y = 120$ 10. $w = 7$

11. $x = 12$ 12. $p = 5$

13. $d = 21$ 14. $h = 3$

15. $k = 35$ 16. $e = 22$

17. $a = 8$ 18. $n = 8$

19. 13 hours

20. $\dfrac{h}{5} = 13$; $h = 65$ inches

Practice C

1. $n = 9{,}180$ 2. $m = 234$

3. $a = 168$ 4. $g = 55$

5. $b = 7$ 6. $e = 15$

7. $145w = 725$; $w = 5$

8. $f \div 21 = 14$; $f = 294$

9. $23b = 253$; $b = 11$

10. $k \div 15 = 47$; $k = 705$

11. $12n = 84$; $n = 7$

12. $m \div 10 = 54$; $m = 540$

13. 48 tons of ore   14. $14,655

15. 78 students   16. 5 pounds

Review for Mastery

1. $x = 18$ 2. $s = 64$

3. $c = 70$ 4. $n = 36$

5. $w = 8$ 6. $b = 6$

7. $z = 5$ 8. $m = 4$

Challenge

1. 16; 4   2. 6; 3

3. 2; 1   4. 9; 3

5. 8; 0   6. 3; 1

        Holt McDougal Mathematics

7. 15; 6; 9   8. 5; 10; 3

9. 3 5 0

10. 2 8 1, 4 2 1, 9 0 6

Problem Solving

1. 11,383 ships

2. $25y = 3,302.75$; $y = 132.11$;
    132.11 yen per dollar

3. $12 = 3y$; $y = 4$; 4 years

4. 24 hours   5. B

6. H   7. C

8. F

Reading Strategies

1. division

2. $\dfrac{108}{9} = \dfrac{9y}{9}$ 3. $y = 12$

Puzzles, Twisters & Teasers

1. $a = 100$ 2. $w = 1$

3. $r = 40$ 4. $h = 32$

5. $d = 9$ 6. $y = 48$

7. $n = 136$ 8. $o = 5$

9. $e = 21$ 10. $u = 27$

11. $b = 37$ 12. $m = 12$

13. $t = 50$ 14. $s = 11$

WHY YOU MUST BE UNDER THE
    WEATHER

          Holt McDougal Mathematics